Principles, Evaluation and Methods in Food Science

NIPA® GENX ELECTRONIC RESOURCES & SOLUTIONS P. LTD.
New Delhi-110 034

Principles, Evaluation and Methods in Food Science

S. Vijaya Jyothi
Associate Professor
Department of Food Science and Nutrition
S.P. Mahila University
Tirupathi-517502, Andhra Pradesh

NIPA® GENX ELECTRONIC RESOURCES & SOLUTIONS P. LTD.
New Delhi-110 034

NIPA® GENX ELECTRONIC RESOURCES & SOLUTIONS P. LTD.

101,103, Vikas Surya Plaza, CU Block
L.S.C.Market, Pitam Pura, New Delhi-110 034
Ph : +91 11 27341616, 27341717, 27341718
E-mail: newindiapublishingagency@gmail.com
Website: www.nipabooks.com

For customer assistance, please contact
Phone: + 91-11-27 34 17 17
Fax: + 91-11-27 34 16 16
E-Mail: feedbacks@nipabooks.com

ISBN: 978-93-91383-76-3

Composed and Designed by NIPA®.

Preface

This laboratory manual has two purposes.

The first purpose is to describe what food science is and what food scientists do.

The second purpose is to describe laboratory experiments that demonstrate practical applications of food science.

Food science is all of the science involved in taking agricultural food products from the farmer's gate to the grocery store, restaurant, or dinner table. Food scientists work with all sectors of agriculture. Food science includes both basic and applied biology, microbiology, chemistry, math, business, engineering, physics, and other disciplines. A food scientist's goal is to make safe, high quality food products that are profitable to all segments of agriculture.

Food science students can also compete in national competitions dealing with food, such as Dairy Judging, Meats Judging, New Product Development, and the Research Chef's Association. These events offer participants networking and learning opportunities for future career growth.

But food science graduates who do not pursue professional degrees in healthcare also have great job opportunities and often advance rapidly. They can work for regulatory agencies, ingredient and equipment manufacturers, research firms, suppliers, or other companies; all offer great opportunities for food science graduates.

Laboratory exercises in this manual demonstrate principles and evaluation practices behind starch, fruits and vegetables, milk and milk products, egg foams and emulsions, meat, fats and oils, sugar and jiggery cookeries. And preservation practicals include drying and dehydration of fruits and vegetables, jams and jelly making, fruit juices, squashes and pickling. These laboratory experiments demonstrate some simple scientific principles that apply to food manufacturing and show the characteristics of some common foods.

Authors

Contents

1

Starch Cookery

1. Gelatinization Properties of Starch

Aim

1. To demonstrate the properties of starches and factors affecting gelation
2. Comparison of properties of different sources of starch eg: viscosity, gelation, and retro gradation

Table 1. Amy lose and Amyl pectin content of cereal root starches (book values)

Source of starch	Amy lose %	Amyl pectin %
Rice (Normal)	25	75
Rice, glutinous	20-Oct	80-90
Wheat	24	76
Corn	28	72
Sorghum	22	78
Waxy rice	2-Jan	98-99
Waxy corn	3-Jan	97-99
Waxy sorghum	3-Jan	97-99
Tapioca	17	83
Potato	22	78
Sago palm	27	73
Banana	21	79
Arrow root	18	82

Cooking quality of cereal and root starches

The stiffness of starch gels from different starches at 5% concentration was shown in decreasing order of stiffness in the following table. Of the cereal starches corn starch gel was firmest followed by wheat and rice starch gels. Among the roots and tubers starches, potato yielded a ropy paste, arrowroot a soft paste while cassava starch gave viscous fluid.

Experiment

Procedure: Ingredients; Basic recipe - starch - 5g water - 95 ml

Mix both starch and water to a starch slurry solution without forming lumps. Then heat the slurry solution on the stove and cook it until 80-90°C till the starch becomes paste. Assess the gel characteristics of each variety of starch by studying the book values and record the observations of different starches taken.

Table 2. Comparative stiffness of cooked starch (5% concentration) from different starches

Starch	Consistency of starch paste (book values)	Record observations Consistency of starch paste	
		Type of starch taken	Type of gel formed
Corn	Stiff Gel	1	
Wheat	Stiff Gel	2	
Rice	Stiff Gel	3	
Potato	Ropy paste	4	
Arrow root	Soft paste	5	
Cassava	Viscous fluid	6	
		7	

2. Comparison of properties of different sources of starch: gelatinization, retro- gradation

Experiment

Procedure: Ingredients; starch- 5g water- 95ml

Mix both the ingredients to slurry without forming lumps. Then heat the slurry on the stove and cook it until 80°C -90°C till the starch becomes paste. Pour the gelatinized mass into two cups. One cup starch paste use for measurement of hot paste viscosity, and second cup use to measure cold paste viscosity (retro gradation) by keep the samples in refrigerator for 1 hour.

Viscosity measurement by Line spread test

Assess the viscosity of starch gels (hot and cold) by pouring one table spoon of gelatinized starch in the centre of line spread sheet and wait for one minute. Observe the spreading of starch solution on four sides A B C D and record in centimeters. Repeat the same measurement for retrograded cold starch paste. Calculate the viscosity value by taking average of 4 values and record % viscosity.

$$\frac{A + B + C + D}{4} = \text{Average value x100} = \text{\% viscosity}$$

% Gel Sag: 1 Take a broom stick and place at the centre of one full cup of gelatinized paste and then place it on a measuring scale to measure the length of the one full cup of gelatinized mass (R1)

2. Repeat the same after the starch cup is inverted on a saucer (R2).

Calculation:

$$\% \text{ gel Sag} = \frac{\text{R1-R2}}{\text{R2}} \times 100$$

Table 3. Comparison of properties of different sources of starch: gelatinization, retro- gradation and % Gel Sag

Type of starch	Temperature at which gelatini-zation completes (°C)	Time taken for gelatini-zation (minutes)	Hot paste viscosity (Line spread test) (Cms) (Cms)	Cold paste viscosity (retro-gradation) (Line spread test)	% Gel Sag	Infer-ences
Corn starch :						
Wheat starch :						
Rice starch :						
Waxy rice :						
Potato :						
Tapioca :						
Sago :						
Arrow root :						
Any other :						

Starch cookery 2

Factors affecting the gelatinization properties of starch

1. Temperature of gelatinization

The temperature at which gelatinization begins varies from about 65-72°C with different starches and gelatinization is complete at about 90°C. It is essential to heat starch to till 90°C for obtaining maximum gelatinization.

Experiment

Procedure; Ingredients; starch- 5g, water- 95ml

Mix both the ingredients to slurry without forming lumps. Then heat the slurry on the stove and cook it. As the cooking proceeds remove a sample of cooked starch when the temperature reaches to 50°C, 60°C, 70°C, 80°C, 90°C and 100°C. Assess the gel quality characteristics as-Very thin liquid/ thin liquid/Thick liquid/Viscous liquid/Thin gels/Thick gels/Hard gels/Very hard gels and record the observations of given starches.

Table 1. Factors affecting temperature of gelatinization

Type of starch	Temperature of cooking					
	50°C/gel quality	60°C/gel quality	70°C/gel quality	80°C/gel quality	90°C/gel quality	100°C/gel quality
Corn starch:						
Wheat starch:						
Rice starch:						
Waxy rice:						
Potato:						
Tapioca:						
Sago:						
Arrow root:						

Demonstration

Prepare vegetable Corn soup using starch and study the effect of Starch on gelatinization/thickening property of corn soup preparation and evaluate it

____

____

____

____

____

2. Factors affecting concentration of starch

Experiment

Procedure: Ingredients; starch - 5g, water - 95ml

Prepare the basic starch solution as well as variations given in the table and cook the starch till it becomes gelatinized and record the observations

Table 2. Effect of concentration of starch on strength of the starch gels

Starch Concentration	Gelatinization temperature ^0C	Time taken for gel formation (minutes)	Gel strength/ weak/strong/ very hard gels	% gel sag	moldable/ not-moldable
Basic starch : slurry	....	....	....	....	....
Basic + 5 g : extra starch	....	....	....	....	....
Basic + 7 g : extra starch	....	....	....	....	....
Basic + 10 g : extra starch	....	....	....	....	....
Basic + 15 g: extra starch	....	....	....	....	....

Demonstration

Prepare thin **consistency Ragi malt** and **thick consistency Ragi malt** using ragi flour as thickening agent and study the effect of addition of extra amount of ragi flour on gelatinization/ thickening property of starch and evaluate it

____

____

____

____

3. Effect of sugar concentration on strength of the starch gels

Experiment

Procedure: Basic recipe: 5g corn starch +95g water

Prepare starch gel made from 5g of corn, wheat or rice starch and 95 g of water (was stiff enough to retain the shape of the mould). When increasing amounts of sugar 10, 30, 50 or 60g sugar added to the basic recipe starch solutions of different starches , the starch gels shows that increasing transparency and tenderness. Record the observations.

Table 3. Effect of sugar concentration on strength of the starch gels and gel properties

Sugar concentration	Gel formation temperature^{0}C	Time taken for gel formation (minutes)	Type of gel setting weak/ medium/ strong	Consistency of gels- tender gels/ thick gels/ transparent	Moldable/ not- moldable
5 g corn starch : + 95 g water	....	....	....	....	....
5 g corn starch : + 95 g water + 10 g sugar	....	....	....	....	....
5 g corn starch : +95 g water + 30 g sugar	....	....	....	....	....
5 g corn starch : +95 g water + 50 g sugar	....	....	....	....	....
5 g corn starch : +95 g water + 60 g sugar	....	....	....	....	....

Demonstration

Prepare Sago Payasam and study the effect of Sugar on gelatinization/thickening property of starch and evaluate it

____

____

____

____

4. Effect of milk on strength of the starch gels

Experiment

Procedure: Ingredients: 5 g corn starch + 95 g milk

When corn starch 5 g was added to 95 g of milk and heated to 95^{0}c, raw milk yielded a stiffer gel than pasteurized milk. Between two samples of pasteurized milk, and milk pasteurized at 72^{0}C gave a stronger gel than milk pasteurized at 80^{0}C. Record the observations

Table 4. Effect of milk additions on gelatinization of starch and gel properties

Milk variation	Gel formation temperature°C	Time taken for gel formation (minutes)	Type of gel setting	Line spread viscosity (cm) moldable/ not-moldable
5 g corn starch + : 95 g raw milk	....	....	....	....
5 g corn starch + : 95 g milk (boiled at 72°C)	....	....	....	....
5 g corn starch + : 95 g milk (boiled at 80°C)	....	....	....	....

5. Effect of Acid on strength of the starch gels

Strong acid like Hcl when added to hot starch gel hydrolyses rapidly the starch and the gel liquefied. Weak acids like citric acid are less active and take a long time to hydrolyze the starch. The gel strength is lowered if lemon juice (containing citric acid) is added to the water before heating the mixture for formation of gel.

Experiment

Procedure: Ingredients; starch- 5g, water- 95 ml

Prepare the starch solutions as per the variations given in the table, cook the starch solutions till it becomes gelatinized and record the observations.

Table 5. Effect of Acid concentration on strength of the starch gels and gel properties

Acid variation	Gel formation temperature °C	Time taken for gel formation (minutes)	Type of gel setting	Line spread viscosity (cm)	% Gel Sag	moldable/ not-moldable
Addition of : strong acid HCL to hot starch gel at 90°C	....	....	....	....	....	....
Addition of : lemon juice to	....	....	....	....	....	....

before heating the starch + water mixture and cook it

Addition of lemon juice to after heating the starch + water mixture : ....

Addition of lemon juice to after heating the starch + sugar + water mixture : ....

Demonstration

Prepare tomato soup using starch as thickening agent and study the effect of acid on gelatinization property of starch and evaluate it

6. Effect of dry heat on strength of the starch gels

Dry heat such as toasting the starch in a hot pan leading to browning of starch, breaks down the starch and reduces the gel strength.

Experiment

Procedure: Ingredients; starch-5g water- 95 ml

Toast the starch for each variation and add water prepare the basic starch solution as well as variations given in the table and cook the starch till it becomes gelatinized and record the observations

Table 6. Effect of dry heat on raw starch on strength of the starch gels and gel properties

Dry heat variation	Gel formation temperature°C	Time taken for gel formation (minutes)	Type of gel setting	Line spread viscosity (cm)	moldable/ not-moldable
Basic gel with : raw starch					
Gel preparation : with mild toasted starch					
Gel preparation : with moderate toasted starch					

7. Effect of enzymes on strength of the starch gels

If amylases are added to the water containing the starch suspension and then heated gradually to 90°c for the formation of gel, then the gel strength is reduced as amylases breaks down the starch molecules rapidly.

Procedure: Ingredients: 5 g corn starch +95 g water

Prepare the basic starch solution as well as variations given in the table and cook the starch till it becomes gelatinized and record the observations.

Table 7. Effect of enzymes on strength of the starch gels and gel properties

Effect of enzyme	Gel formation temperature°C	Time taken for gel formation (minutes)	Type of gel setting	Line spread viscosity (cm)	moldable/ not-moldable
5g corn starch + : 95g water					
5g corn starch + : 95 g water + amylase enzyme					

2

Leavening Agents

Aim: To demonstrate the role of leavening agents in fermented foods like Idly, dosa and bakery foods like bread, cakes and biscuits

Principle

Leavening agents produce carbon dioxide which helps to give characteristic structure to bread cakes and biscuits etc. In the case of cakes, leavening is produced by air entrapped in the dough due to the presence of egg in the mix. The important groups of leavening agent's are l. baker's yeast 2. Chemical leavening agents

Baker's yeast: Yeast used in baking may be in the form of moist pressed cake or in the form of dehydrated granules/active dry yeast cells. Yeast cells when added to the dough, will ferment sugar and produce CO_2 and alcohol. This fermentation is gradual increases in rate with increase in time and temperature. The distribution of CO_2 in the elastic dough prior to baking is responsible for the characteristic structure of bread after baking.

Chemical leavening agents

1. Baking powder
2. Baking Soda

1. Demonstration on role of leavening agents in fermented foods

Idly preparation

Experiment

Procedure: Ingredients: rice-2 cups black gram dhal-1 cup

Soak the ingredients as per variations given in the table 1. After soaking drain the water and wash the ingredients and put for grinding in a wet grinder/mixer till rice obtained as a coarse paste and black gram dhal as a soft and fluffy mass. Combine both the pastes and add little salt mix thoroughly and keep it for overnight fermentation. After fermentation measure the batter for batter properties

1. **Aeration of the batter:** Microscope testing of batter for air bubbles
2. **Volume of the batter (ml)**: Measuring cylinder

   Prepare idly/dosa as per the variations given in the table 2. Record the effect of leavening agents in idly variations

Table 1. Effect of soaking variations on fermentation capacity of the idly batter

Variation soaking in	Raw/ parboiled rice weight (gms) before soaking	Raw/ parboiled rice weight (gms) after soaking	Black gram dhal Weight (gms) before soaking	Black gram dhal Weight (gms) after soaking	Aeration of the batter before Fermentation	Aeration of the batter after Fermentation
3 hrs of soaking	2 cups (100)	....	1 cup (50)	....	....	....
6 hrs of soaking	2 cups (100)	....	1 cup (50)	....	....	....
9 hrs of soaking	2 cups (100)	....	1 cup (50)	....	....	....
12 hrs of soaking	2 cups (100)	....	1 cup (50)	....	....	....

Table 2. Effect of fermentation and added leavening agents on quality characteristics of idly's

Leavening agents variation	Volume of the batter (ml) after fermentation	Weight of idly's(gms)	Softness of idly's	Taste/flavor of idly's	Color of idly's	Porous texture of idly cells
Without Baking soda	(3 hrs of soaking Batter) =	....	....	....	....	....
	(6 hrs of soaking Batter) =	....	....	....	....	....
	(9 hrs of soaking Batter) =	....	....	....	....	....
	(12 hrs of soaking Batter) =	....	....	....	....	....
With Baking soda	(3 hrs of soaking Batter) =	....	....	....	....	....
	(6 hrs of soaking Batter) =	....	....	....	....	....
	(9 hrs of soaking Batter) =	....	....	....	....	....
	(12 hrs of soaking Batter) =	....	....	....	....	....

2. Demonstration on role of leavening agents in bakery products

Bread preparation

Procedure: Weigh the ingredients according to the formulae given in table and prepare the bread as per the procedure given and evaluate the characteristics.

Table 3. Proportions of ingredients used in bread formulations

Formulae for different types of breads

Ingredients	White bread-I	Whole wheat bread	Gluten free bread
Flour	100	20	60
Whole wheat flour	—	80	——
Millet flour	—	——	20
Flax seed meal	—	——	20
Water	60.6	60.6 + as needed	60.6 + as needed
Yeast	1.8	1.8	1.8
Salt	2	2	2
Milk dry (skim)	5	5	5

Method of bread preparation

1. Mixing of ingredients and preparation and kneading of dough till it develops smooth surface and rest for 15-20 minutes so that gluten develops well
2. Fermentation of the dough for 1 hour
3. Baking of bread dough; bread is baked at 226.7°C for about 30 min. as the dough enters the hot oven maintained at 226.7°C, a visible film forms on the dough surface. Next, the expansion of the dough volume to the extent of 30% occurs.
4. Cooling and packing; cooling of bread is essential in order to avoid moisture condensation before packing. Packing of bread should be done in moisture proof films or wrapping in waxed paper to prevent loss of moisture from crust and crumb.

Table 4. Evaluation of bread preparation

Bread variation	Aeration of dough before fermentation (volume of dough) ml	Aeration of dough after fermentation (volume of dough) ml	Yield of bread after baking cooked well/ or under cooked	Crust structure character- istics Thick, dry/ or soft, moist	Crumb structure character istics Soft & spong /or sticky crumb
Bread made with Std. yeast	....	....	....	....	....
Bread made with less yeast	....	....	....	....	....

Bread made	....	....	....	....	....
with more	....	....	....	....	....
yeast	....	....	....	....	....
Bread made	....	....	....	....	....
with whole	....	....	....	....	....
wheat flour	....	....	....	....	....
with std.	....	....	....	....	....
yeast	....	....	....	....	....
Bread made	....	....	....	....	....
without	....	....	....	....	....
yeast	....	....	....	....	....

Role of leavening agents in Cakes

Procedure: Weigh the ingredients according to the formulae given in table and prepare the cakes as per the procedure given and evaluate the characteristics.

Table 5. Formulae for different types of sponge cakes

Angel cake		Sponge cake	
Ingredients	**Amount g**	**Ingredients**	**Amount g**
Cake flour	90 g	Cake flour	100 g
Sugar	250 g	Sugar	200 g
Cream of tartar	4.0 g	Egg	300 g
salt	1.0 g	Lemon juice	15.0 g
egg white	250 g	Water	30 g
water	30.0 g	Lemon rind grated	1 table spoon
		Salt	1.0 g

Cake preparation

1. Sugar is mixed well with fat, and then beaten whole egg is added to the sugar-fat mixture and mixed well with uniform consistency and fluffy mass (fat-sugar-egg blend).
2. Sift/sieve the flour along with baking powder and salt.
3. Milk and flavorings are combined. Small quantities of flour and milk are combined and added alternatively to the fat-sugar-egg blend. The blend should be mixed after each addition.
4. Mix the blend well till a uniform batter of pouring consistency is obtained.
5. Baking: once the batter is prepared the batter is transferred to cake pans. The pans are kept in a baking oven set at 190^0c and baked for 30 minutes.

Table 6. evaluation of cake preparation

Cake variation	Creaming performance-fluffy/thick/thin mass	Consistency of batter before baking thick/thin	Volume of cake after baking	Well cooked/ under cooked	Crust structure characteristics	Crumb structure characteristics
Basic cake/ std. cake	....	....	....	....	....	....
Basic cake without baking powder	....	....	....	....	....	....
Basic cake with less baking powder	....	....	....	....	....	....
Basic cake with more baking powder	....	....	....	....	....	....
Basic cake without eggs	....	....	....	....	....	....
Basic cake with less eggs	....	....	....	....	....	....
Basic cake with more eggs	....	....	....	....	....	....
Basic cake with insufficient creaming step	....	....	....	....	....	....

Role of leavening agents in Biscuits

Procedure: weigh the ingredients according to the formulae given in table and prepare the biscuits as per the procedure given and evaluate the characteristics.

Table 7. Proportions of ingredients used in biscuit formulations

Ingredients	Sweet biscuits	Salt biscuits
Soft wheat flour	100	100
Shortening (fat)	10	10
Cane sugar	20	5
Salt	-	0.6
Sodium bicarbonate	0.4	0.4
Ammonium carbonate	0.2	0.2
Flavor	0.1	1.2

Steps in biscuit preparation

Mixing and kneading; dry ingredients like flour and sugar are transferred to a mixing vessel, whereas shortenings and flavorings are added and mixed in a mechanical mixer. Then water and baking powder are added to mixer and continue mixing till a desired consistency of dough is obtained.

Sheeting and shaping; the dough is rolled into sheets of desired thickness

The rolled dough sheet is cut into different shapes and size using biscuit moulds/ cutters.

Baking and cooling; the biscuits are baked at 232.2^{0}C for 15 min and cool the biscuits

Packaging; the biscuits should be packed in tins or moisture proof packaging. The moisture proof and grease proof glassine, cellophane and metal foil laminated moisture proof papers extend the shelf life of biscuits for long periods for 6 months without the development of off-flavors.

Table 8. evaluation of biscuits preparation

Biscuit variation	Baking performance- able to spread the biscuit/not- able to spread	Diameter of biscuit (cm)	Brittleness of biscuit-crispy/ crunchy/	Dry or moistness biscuit
Standard biscuits recipe	....	....	....	....
Standard biscuits with baking powder + baking soda	....	....	....	....

Standard biscuits	....	....	....	....
with baking	....	....	....	....
soda only	....	....	....	....
Standard biscuits	....	....	....	....
without baking	....	....	....	....
powder and	....	....	....	....
baking soda	....	....	....	....

3

Pulse Cookery

Factors effecting Pulse cookery

Aim: To demonstrate cooking time of pulses with soaked and un-soaked pulses

Procedure: Weigh the pulses: 50 g, water: add sufficiently

1. Clean and wash the pulses, Mix the pulses with water, soak for 1 hour, 3 hours, 6hours, 9hours and overnight.
2. After soaking drain the water wash thoroughly and add fresh water keep it for boiling/cooking.
3. Study the effect of Acid and effect of Alkali on cooking time of pulses.

   While cooking note down the cooking time and cooking quality characteristics of pulses given and record the observations

Table 1. Evaluation of cooking time of pulses with soaked and un-soaked pulses

Soaking variation	**Red gram dhal**		**Whole green gram**		**Whole Bengal gram**		**Peas**		**Peas**	
	Cooking time (min)	**Cooking quality/ soft/ semi hard/hard**	**Cooking time (min)**	**Cooking quality/soft/ semi hard/ hard**	**Cooking time (min)**	**Cooking qualit/ hard/ hard y**	**Cooking time (min)**	**Cooking quality/ soft/semi hard/hard**	**Cooking time (min)**	**Cooking quality**
Pulses without soaking										
Pulses with 1hr soaking										
Pulses with 3hr soaking										
Pulses with 6hr soaking										
Pulses with 9hr soaking										
Pulses with overnight soaking										

Table 2. Evaluation of cooking time of pulses with addition of acid and alkali

Variation	**Red gram dhal**		**Whole green gram**		**Whole Bengal gram**		**Peas**		**Peas**	
	Cooking time (min)	**Cooking quality/ soft/semi hard/hard**	**Cooking time (min)**	**Cooking quality/soft/ semi hard/hard**	**Cooking time (min)**	**Cooking qualit/ hard/ hard y**	**Cooking time (min)**	**Cooking quality/ soft/semi hard/hard**	**Cooking time (min)**	**Cooking quality**
Cooking pulses with addition of citric acid/ dil.lemon juice/vinegar										
Cooking pulses with addition of alkali/soda										

4

Vegetable Cookery

Factors effecting Vegetable cookery

Aim: To study the effect of cooking time, cooking media, cooking methods, metal ions and on color, texture and flavor and palatability of different vegetable and fruits

Effect of Cooking Time on color, texture and flavor of vegetables

Procedure: weigh vegetables 20 g (for each variation) Water: add sufficiently

1. Cook the vegetable in water in a pan according to the "cooking time" variations given in the table 1 and note down the color, texture and flavor changes takes place during cooking for each variation and record the observations:

Table 1. Effect of Cooking Time on color, texture and flavor changes of given vegetables

Cooking time variation (minutes)	Green leafy veg			Beet root			Carrot			Beans		
	Color	Text-ure	Fla-vor	Color	Text-ure	Fla-vor	Color	Text-ure	Fla-vor	Color	text-ure	Fla-vor
5 min	....	....	....	....	....	....	....	....	....	....	......	....
10 min	....	....	....	....	....	....	....	....	....	....	......	....
15 min	....	....	....	....	....	....	....	....	....	....	......	....
30 min	....	....	....	....	....	....	....	....	....	....	......	....
45 min	....	....	....	....	....	....	....	....	....	....	......	....
60 min	....	....	....	....	....	....	....	....	....	....	......	....

Effect of Cooking Time on odor/flavor of strong flavored vegetables

Procedure: Weigh vegetables 20 g (for each variation) Water: add sufficiently

1. Cook the vegetable in water in a pan according to the "cooking time" and "cooking method" variations given in the table 2.
2. Collect the cooking water for each variation in the test tubes and add lead acetate for the cloudy appearance of the solution in the test tube denotes **the presence of H_2S**
3. After tested with lead acetate for all the test tubes solutions, record the observations for each variation **for the presence of H_2S**

Table 2. Effect of Cooking Time on odor/flavor of strong flavored vegetables

Cooking method variation	Presence of H_2S in cooking water tested with lead acetate in a test tube								
	Onion 10 min cooking/ cooking water	Onion 20 min cooking/ cooking water	Cabbage- 10 min cooking/ cooking water	Cabbage- 20 min cooking/ cooking water	Cabbage- 30 min cooking/ cooking water	Cauliflower- 10 min cooking/ cooking water	Cauliflower- 20 min cooking/ cooking water	Garlic 10 min cooking/ cooking water	Garlic 20 min cooking/ cooking water
Open vessel cooking	....	....	....	....	....	....	....	....	....
Partial open lid cooking	....	....	....	....	....	....	....	....	....
Cooking with closed lid	....	....	....	....	....	....	....	....	....
Pressure	....	....	....	....	....	....	....	....	....
Cooking	....	....	....	....	....	....	....	....	....

Effect of cooking media on color, texture and flavor of vegetables

Procedure: Weigh vegetables 20 g (for each variation) Water: add sufficiently

1. Cook the vegetable in water in a pan according to the "cooking media" variations given in the table 3 and note down the color, texture and flavor changes takes place during cooking for each variation and record the observations:

Table 3. Effect of cooking media on color, texture and flavor of given vegetables

Cooking time variation	Green leafy veg			Beet root			Carrot			Beans		
	Color	Text-ure	Fla-vor	Color	Text-ure	Fla-vor	Color	Text-ure	Fla-vor	Color	text-ure	Fla-vor
Cooking in citric acid media	....	....	....	....	....	....	....	....	....	....	....	....
Cooking in Alkali media	....	....	....	....	....	....	....	....	....	....	....	....
Cooking 60 min in iron kadai	....	....	....	....	....	....	....	....	....	....	....	....
Cooking 60 min in aluminum vessel	....	....	....	....	....	....	....	....	....	....	....	....
Cooking 60 min in stainless steel vessel	....	....	....	....	....	....	....	....	....	....	....	....
Cooking vegetables cut with iron-rust knife	....	....	....	....	....	....	....	....	....	....	....	....

Effect of cooking methods on color, texture and flavor of vegetables

Procedure: weigh vegetables 20 g (for each variation) Water: add sufficiently

1. Cook the vegetable in water in a pan according to the "cooking method" variations given in the table 4 and note down the color, texture and flavor changes takes place during cooking for each variation and record the observations:

Table 4. Effect of cooking methods on color, texture and flavor of given vegetables

Cooking time variation	Beans			Green leafy veg			Carrot			Potato		
	Color	Texture	Flavor	Color	Texture	Flavor	Color	Texture	Flavor	Color	texture	Flavor
Open vessel	....	....	....	....	....	....	....	....	....	....	......	....
cooking	....	....	....	....	....	....	....	....	....	....	......	....
Partial open	....	....	....	....	....	....	....	....	....	....	......	....
lid cooking	....	....	....	....	....	....	....	....	....	....	......	....
Cooking	....	....	....	....	....	....	....	....	....	....	......	....
with closed	....	....	....	....	....	....	....	....	....	....	......	....
lid	....	....	....	....	....	....	....	....	....	....	......	....
Pressure	....	....	....	....	....	....	....	....	....	....	......	....
cooking	....	....	....	....	....	....	....	....	....	....	......	....
Cooking on	....	....	....	....	....	....	....	....	....	....	......	....
open fire	....	....	....	....	....	....	....	....	....	....	......	....
wood	....	....	....	....	....	....	....	....	....	....	......	....
Steam	....	....	....	....	....	....	....	....	....	....	......	....
cooking	....	....	....	....	....	....	....	....	....	....	......	....
Microwave	....	....	....	....	....	....	....	....	....	....	......	....
cooking	....	....	....	....	....	....	....	....	....	....	......	....
Roasting	....	....	....	....	....	....	....	....	....	....	......	....
Baking	....	....	....	....	....	....	....	....	....	....	......	....

Stewing	....	....	....	....	....	....	....	....	....	....	......	....
Shallow Frying	....	....	....	....	....	....	....	....	....	....	....	....
Deep frying	....	....	....	....	....	....	....	....	....	....	....	....

Vegetable cookery 2

Browning reactions in fruits and vegetables

Aim: To learn the browning reactions in fruits and vegetables and to study the preventive measures of browning

Principle of enzymatic browning: When the vegetable tissue is cut or injured, the cut surface is exposed to air, followed by phenol oxidase enzymes released at the surface act on polyphenols present in tissue, oxidizing them to orthoquinones.

The rapid darkening of the cut surface of apple, banana, brinjal(egg plant), plantain and potato are examples of enzymatic browning due to presence of natural phenolic compounds present in intact tissues.

Procedure: Ingredients: apple, banana, brinjal (egg plant), plantain and potato

1. Clean and wash the vegetables and fruits
2. Cut them into ideal size
3. Prepare the browning reaction preventive solutions/or solids/ or powders according to the methods given in the table 1
4. Add the cut fruit/ or vegetable pieces in to each of the solutions/ sprinkle powders/sprinkle granules etc
5. Note down the presence of browning as positive (+)
6. Note down the absence of browning as Negative (-)
7. Record the observations for all variations in table 1

Table 1. Prevention of enzymatic browning in fruits and vegetables

Preventive method	Apple	Banana	Brinjal (egg plant)	Plantain	Potato
Thermal inactivation					
1. Blanching (boil veg.	....	....	....	....	....
in hot water at 100^{0}C	....	....	....	....	....
for 2-10 minutes	....	....	....	....	....
2. Change of pH using acids	....	....	....	....	....
Soaking cut F/V in	....	....	....	....	....
6-7 pH solution	....	....	....	....	....
Soaking cut F/V in	....	....	....	....	....
4.0 pH solution	....	....	....	....	....
Effect of antioxidants					
1. Sprinkle ascorbic	....	....	....	....	....
acid on fruit surfaces	....	....	....	....	....
2. Sprinkle ascorbic	....	....	....	....	....
acid and citric acid	....	....	....	....	....
on fruit cut surfaces	....	....	....	....	....
3. Sprinkle sulphur dioxide/	....	....	....	....	....
sulphites/ and bisulphites	....	....	....	....	....
on vegetable cut surfaces	....	....	....	....	....
Prevention of contact with oxygen					
1. Keep in refrigerator	....	....	....	....	....
2. Tie in wet muslin cloth	....	....	....	....	....
3. Keep in salt solution	....	....	....	....	....
4. Keep in sugar solution	....	....	....	....	....
5. Sprinkle salt on cut surfaces	....	....	....	....	....
6. Sprinkle sugar on cut surfaces	....	....	....	....	....
7. Heating cut f/v in live steam	....	....	....	....	....

Demonstration

Demonstrate retention of color pigments in any vegetable recipes using above methods and evaluate it

___

___

___

___

5

Milk Cookery

Physic-chemical properties of Milk

Aim: To study the physicochemical properties, effect of heat, acid, enzymes and effect of cooking on milk proteins

Physical Properties of milk

Procedure: Collect different types of market milk samples and study the properties of milk. Boil the milk and note down the properties given in table 1 and record the observations

Table 1. Physicochemical Properties of milk

Milk types	Color	Boiling point	pH	Density	Skin characteristics
Cow's Pure milk	....	....	....	....	....
Buffalo's pure milk	....	....	....	....	....
Diluted milk	....	....	....	....	....
Homogenized milk (packet milk)	....	....	....	....	....
Skim milk	....	....	....	....	....
Condensed/evaporated milk	....	....	....	....	....
Soya milk	....	....	....	....	....
Toned or double toned milk	....	....	....	....	....

Coagulation properties of milk proteins

Procedure:

1. Collect market milk samples 2. Collect the acidic agents like lime, citric acid, rennin enzyme
2. Study the properties of milk by addition of acidic agents separately to the boiling milk and note down the **coagulation properties** according to table 2 and record the observations

Table 2. Coagulation properties of milk proteins

Variation in milk coagulants	Coagulated / not coagulated	Formation of Thin curd / thick curd
Milk without coagulants	....	....
Milk with lime juice	....	....
Milk with citric acid	....	....
Milk with rennin enzyme	....	....
Milk with citric acid + rennin enzyme	....	....

Coagulation property of milk proteins upon cooking vegetables with milk

Procedure:

1. Collect market milk samples
2. Collect the vegetables
    - Clean and cut the vegetables
    - Take a sauce pan adds vegetables and water according to the variation given in table 3
    - Start cooking the vegetable
    - When half cooking is done add boiled milk and cook till vegetable becomes soft
    - Note down the coagulation characteristics of milk cooking in each variety of vegetable
    - Record the observations

Table 3. Coagulation property of milk proteins upon cooking vegetables with milk

Cooking of vegetables with milk	Milk coagulated	Milk not coagulated
Cooking of carrots with milk	....	....
Cooking of beans with milk	....	....
Cooking of ridge gourd with milk	....	....
Cooking of bottle gourd with Milk	....	....
Cooking any other vegetable with milk	....	....

Prepare Curds using different types of market samples and evaluate the curd setting properties

Table evaluation of Curd setting properties

Type of milk used for curd setting	Body texture Loose/Semi solid/Solid/ Firm/Hard	Color/flavor	Sourness	Taste
Cow's Pure milk	....	....	....	....
Buffalo's pure milk	....	....	....	....
Diluted milk	....	....	....	....
Homogenized milk (packet milk)	....	....	....	....
Skim milk	....	....	....	....
Condensed/ evaporated milk	....	....	....	....
Soya milk	....	....	....	....
Toned or double toned milk	....	....	....	....

Aim to study the properties of milk in the preparation of cream of tomato soup and curd setting when curd cultured at different conditions and Cheese preparation.

1. Preparation and evaluation of cream of tomato soup

Procedure

- Prepare white sauce using white flour and butter
- Boil the milk
- Prepare pure tomato juice

Prepare cream of tomato soup by mixing the white sauce, milk and tomato juice according to the following proportions and variations given in table 1 and note down the characteristics and record the observations

Table 1. Preparation and evaluation of cream of tomato soup

Cream of tomato soupvariations	Soup characteristics	Taste/flavor	Curdling is observed/ or not observed
Prepare basic tomato soup	....	....	....
Transfer hot milk to hot	....	....	....
tomato juice	....	....	....
Transfer hot tomato juice	....	....	....
to hot milk	....	....	....
Add white sauce to hot milk	....	....	....
& transfer to hot tomato juice	....	....	....
Add white sauce to hot	....	....	....
tomato juice and transfer	....	....	....
to hot milk	....	....	....
Add white sauce to cold milk	....	....	....
& transfer to hot tomato juice	....	....	....
Add white sauce to cold	....	....	....
tomato juice and transfer	....	....	....
to hot milk	....	....	....
Add white sauce to cold	....	....	....
tomato juice and transfer	....	....	....
to cold milk	....	....	....

2. Preparation and evaluation of curds cultured at different conditions

Procedure

1. Collect market milk samples
2. Sour curd for culturing

Boil the milk and add the culture according to the conditions given in table 2 and keep the cultured milk over night for appropriate curd setting. Next day note down the curd characteristics and record the observations

Table 2. Evaluation of curd characteristics cultured at different conditions

Curd cultured at different variations	Type of curd setting- thick/thin/ watery	Curd Texture Smooth and soft/ rubbery/soggy/ ropy/loosen	Curd flavor Sweet/sour/ buttery/off-flavor/fungus flavor	Curd color White/cream/ yellow/ greenish /any other
Addition of culture in milk at look warm temperature	....	....	....	....
Addition of culture in hot boiled milk	....	....	....	....
Addition of culture in milk at 75°C	....	....	....	....
Addition of culture in milk at 50°C	....	....	....	....
Addition of culture in milk at room temperature	....	....	....	....
Addition of culture in cold milk	....	....	....	....
Addition of culture to raw milk	....	....	....	....

3. Preparation and evaluation of Cheese

Procedure: Ingredients: 1. Milk 2. Citric/lemon juice 3. Rennet enzyme 4. Beakers of different sizes 5. Muslin cloth 6. Thermometer

1. Prepare coagulated milk using 1000ml. Acidify the milk by adding citric acid/lemon juice and coagulate the milk by addition of rennet enzyme solution (20 ml per 1000 kg of milk).
2. Divide the coagulated milk into 7 portions

3. Place the each portion of coagulated milk in a small beaker
4. Insert the small beaker in large beaker contains small amount of water 5. Place the beaker set on the wire gauge when placed it on stove.
5 Place the thermometer in the centre of the beaker
6. Heat the coagulated milk solution at different temperatures given in (table 1).
7. After reaching the target temperatures remove the beaker and transfer the coagulated solution to another beaker by straining through muslin cloth to remove the whey and squeezed off
8. When curd is formed of proper firmness remove the muslin cloth and place the curd on saucer
9. Evaluate the curd textural characteristics and record the observations

Table 1. Evaluation of textural characteristics cheese

Cheese variation	**Type of cheese formed- granular particles/smooth uniformity body/ watery**	**Cheese texture soft and smooth/ semi hard and rough/hard and rubbery**	**Cheese flavor Bland/sweet/ off-flavor**	**Cheese color White/cream/ yellow/greenish/ any other**
Heat coagulated milk till 40°C	....	....	....	....
Heat coagulated milk till 50°C	....	....	....	....
Heat coagulated milk till 60°C	....	....	....	....
Heat coagulated milk till 70°C	....	....	....	....
Heat coagulated milk till 80°C	....	....	....	....
Heat coagulated milk till 90°C	....	....	....	....
Heat coagulated milk till 100°C	....	....	....	....

6

Preparation of Milk Products

Aim to study the preparation and evaluation of different m**ilk Products**

1. Preparation and evaluation of Ice-Cream

Procedure: Ingredients; milk fat (cream)-12%, non-fat milk solids (skim milk powder)-11%, sugar-15%, stabilizer-0.2%, emulsifier-0.2%, flavors-a pinch

Basic ice cream preparation: 1. Mix the ingredients 2. Pasteurization 3. Homogenization 4. Ageing (allowing to standby 24 hours at 4.44^{0}C) 5. Freezing (the mix is frozen at 5.556^{0}c) 6. The freezing must be quick in order to prevent formation of large crystals. The ice cream is semi solid at this temperature. 7. The ice cream is finally frozen to a hard mass at about -30C.

Table 1. Evaluation of Ice-Cream

Ice-Cream variations	**Setting of ice cream-unsetting/ partial setting/ complete setting**	**Texture Rough/ hard/crystals/ granular type/ smooth body**	**Feeling on palate-Crystals were sensed/ granular/sand particles type**	**Flavor / Color Appealing/not-appealing smooth**
Basic ice cream	....	....	....	....
Ice cream prepare with full cream milk	....	....	....	....
Ice cream prepare with skim milk	....	....	....	....
Ice cream prepare with condensed/ evaporated milk	....	....	....	....
Ice cream prepare with soya milk	....	....	....	....
Ice cream prepare with buffalo milk	....	....	....	....
Ice cream prepare with cow's milk	....	....	....	....
Prepare ice cream without ice cream mix	....	....	....	....
Prepare ice cream without sugar	....	....	....	....
Prepare ice cream without corn flour	....	....	....	....
Prepare ice cream without flavor	....	....	....	....

2. Preparation and evaluation of Butter

Procedure: There are three stages in the preparation of butter from good quality cream:

1. Boil the milk till thick cream is formed
2. Add sour culture to the milk when it is look warm temperature
3. Next day curd setting occurs and cream is ripened
4. When ripened cream is churned, the fat coalesces and forms progressively larger clusters of fat globules
5. The fat finally breaks away from the aqueous phase forming a plastic fat called butter
6. Butter is usually salted and edible color is added and stored

3. Preparation and evaluation of Ghee (Butter oil)

Ghee is prepared by melting butter in a stainless steel vessel removing the moisture present by applying heat to the vessel. When the moisture is completely removed, a pleasant aroma develops and browning of the protein takes place. On cooling, ghee sets as a cream colored fat. The keeping quality of ghee will be adversely affected, if the milk contains excessive amounts of copper derived from the containers.

Table 2. Evaluation of butter and ghee

Type of milk	Quantity of ripened cream obtained	Quantity of butter formed after churning	Butter plasticity character	Flavor & taste	Ghee preparation approx. quantity	Ghee Aroma characteristics
Cow's Pure milk	....	....	....	....	....	....
Buffalo's pure milk	....	....	....	....	....	....
Diluted milk	....	....	....	....	....	....
Homogenized milk (packet milk)	....	....	....	....	....	....
Skim milk	....	....	....	....	....	....

Condensed/	....	....	....	....	....	....
evaporated	....	....	....	....	....	....
milk	....	....	....	....	....	....
Soya milk	....	....	....	....	....	....
Toned or	....	....	....	....	....	....
double toned	....	....	....	....	....	....
milk	....	....	....	....	....	....

7

Egg Cookery

Aim: To study the properties of egg coagulation, clarifying and foaming agent-effect of quality of eggs on these properties, egg as emulsifying agent – preparation and evaluation of mayonnaise.

1. Standards of eggs/ or Grades of eggs

In order to assess the quality characteristics of eggs the book reference values were taken for comparison

Table 1. Standards of eggs/ or grades of eggs (book standards)

Quality factor	AA	A	B	Loss
Air cell	1/8 inch or less in depth	3/16 inch or less but>1/8 inch in depth	More than 3/16 inch in depth	Does not apply
White	Clear Firm	Clear may be reasonably Firm	Clear may be Weak and Watery	Does not apply
Yolk	Outline slightly defined	Outline may be Fairly Well defined	Outline Clearly Visible	Does not apply
Blood or Meat Spot	None	None	Blood or Meat Spots totaling number more than 1/8 inch diameter	Blood or Meat Spots totaling number more than 1/8 inch diameter

2. Testing the grades of egg

Procedure: Collect eggs locally from various shops. Examine the eggs according to the standards given in the table 1 and write their characteristics.

Table 2. Testing the quality characteristics of eggs from local shops

Quality factor	**Eggs from shop 1**	**Eggs from shop 2**	**Eggs from shop 3**	**Eggs from shop 4**
Air Cell	....	....	....	....
White	....	....	....	....
Yolk	....	....	....	....
Blood/ or meat spot	....	....	....	....

3. Coagulation properties of egg proteins

Procedure: Soft cooked eggs; the objective is to obtain a soft coagulated white and a yolk that is semi-liquid. Two satisfactory methods are used to obtain soft boiled or cooked eggs.

Boil eggs to Soft cooked eggs according to the Method I and Method II and record the characteristics and observations

Table 3. Evaluation of Coagulation properties of Soft cooked eggs

Method I	**Results- Method I**	**Method II**	**Results- Method II**
Add eggs to boiling water. Turn off the heat to cover/close the pan. And allow the eggs to remain in the water for 5 to 6 minutes.		Add eggs to water at about simmering temperature 85^{0}c and maintain the temperature for 5 to 6 minutes.	

Procedure: Hard cooked eggs; white of hard cooked eggs should be firmly coagulated and the yolk should be solid and mealy. If the yolk is waxy, it is not fully cooked. Two satisfactory methods are used to obtain hard boiled or cooked eggs.

Boil eggs to Hard cooked eggs according to the Method I and Method II and record the characteristics and observations

Table 4. Evaluation of Coagulation properties of Hard cooked eggs

Method I	Results-Method I	Method II	Results-Method II
Add eggs to boiling water. Turn off the heat to cover/close the pan. And allow the eggs to remain in the water for 45 minutes to 1 hour in a warm place.		Add eggs to water at about simmering temperature 85^{0}c and maintain the temperature for 25 minutes.	

4. Formation of dark greenish discoloration in hard boiled eggs

When egg is cooked in boiling water for 15 minutes or longer a dark greenish color (FeS) may be formed on the surface of egg yolk.

Table 5. Evaluation of dark green color (FeS) in hard boiled eggs

Eggs cooked at temperature ^{0}C	Cooking time	Color of surface of yolk
70	1 hour	
85	30-35 minutes	
100	15 minutes	
100	15 minutes cooking and cooled in cold water	
100	30 minutes cooking	
100	30 minutes cooking and cooled in cold water	

5. Omelets and factors affecting it

Plain omelet: Beat the egg with pepper and salt until the yolk and albumen are thoroughly mixed. The pan is heated and the fat allowed melting. The beaten egg is added and spread evenly with a fork. The omelet is cooked on both sides till it attains a light brown color.

French omelet: Mix both eggs and the liquid and beaten until they form a homogenous mixture. The beaten mixture is poured on a heated pan containing melted fat and the mixture is allowed to coagulate undisturbed. Turn on both sides with a spatula till all the sides of omelet gets cooked.

Puffy (foamy) omelet: Beating the yolk till it is very thick and beat albumen separately until the peaks just bend over and folding them together carefully and transfer to the butter melted pan and heat it for 2 minutes to brown the bottom surface and then baking in a skillet/oven at 163^{0}c for 5 minutes.

Baked omelet or **Soufflés:** Soufflé is a baked omelet with added cream sauce. Cook the white sauce before adding to egg whites. The egg whites are separated and beaten until the peaks are stiff. Beat the egg yolk thoroughly and fold both egg yolk and white sauce into the stiff egg white. After folding all the ingredients bake in oven at a temperature of 176.7^{0}C in a pan of water for about 40-50 minutes. Flavoring materials such as cheese, meat, pureed vegetables are added to the mixture for main dish soufflé. Sugar and chocolate and pureed fruit are added for hot dessert soufflé-prune soufflé.

Table 6. Evaluation of factors affecting omelet quality

Omelet type	Variations	Omelet quality	Omelet color	Moistness/ or Dryness of omelet
Plain omelet	Std. preparation	....	....	....
	Improper beating of yolk and albumen	....	....	....
	Cook the omelet on one side	....	....	....
French omelet	Std. preparation	....	....	....
	Addition of 12 ml of liquid	....	....	....
	Addition of vinegar with water	....	....	....
	Addition of water with cream of tartar	....	....	....
	Addition of tomato juice	....	....	....
	Addition of lemon juice with water	....	....	....

	Addition of milk	....	....	....
Puffy (foamy) omelet	Std. preparation	....	....	....
	Bake the omelet in skillet	....	....	....
	Bake the omelet in oven	....	....	....

6. Clarifying properties of eggs

Method: When a clear soup is desired, a small amount of egg albumin is added to the soup liquid in the cold. When the mixture is heated, egg albumin coagulates and carries all suspended particles along with it. On allowing it to settle, a clear soup or boullion or consommé is obtained.

Table 7. Evaluation of clarifying properties of eggs

Soup/boullion/consomme	Addition of egg albumin	Clear soup is obtained / or Not obtained
Prepare a thin soup by stewing vegetables in water	Add egg albumin	....
Prepare a thin soup by stewing vegetables in water	Do not add egg albumen	....
Prepare a thin soup by stewing meat/fish in water	Add egg albumin	....
Prepare a thin soup by stewing meat/fish in water	Do not add egg albumen	....

7. Foaming properties of eggs

In several products, eggs are used to form foams. In meringue and sponge cakes, the function of egg is to stabilize the foam i.e. air in liquid dispersion.

Egg white Foam; Egg white has a unique property of forming a foam when beaten. This property is used in the preparation of cakes, meringue etc. The factors responsible for formation of stable foam are;

1. Low surface tension
2. Low vapor pressure
3. Tendency for the substance on the surface to solidify. Egg white has all these properties.

Table 8. Evaluation of foaming properties of eggs

Factors affecting the foam formation	Variations (Transfer egg albumen in a bowl & beat it according to variations for foam formation)	Beating time (min)	Drainage collected in measuring cylinder(ml)	Stable foam/ or un-stable foam	Foam characteristics
Effect of beating	Standard beating time	....	....	....	....
	Less than Standard beating time	....	....	....	....
	More than Standard beating time	....	....	....	....
Effect of sugar	Add 5 g sugar	....	....	....	....
	Add 10 g sugar	....	....	....	....
Effect of organic acids	Add 0.2% of cream of tartar	....	....	....	....
	Add 0.8% of cream of tartar	....	....	....	....
Effect of temperature	Beating egg white at 15^{0}C	....	....	....	....
	Beating egg white at 35^{0}C	....	....	....	....
Effect of addition of milk	Addition of milk	....	....	....	....
Effect of addition of water	Addition of water	....	....	....	....
Effect of addition of salt	Addition of salt	....	....	....	....
Effect of addition of fat	Addition of fat	....	....	....	....
Effect of addition of whole egg	Addition of whole egg	....	....	....	....

8. Thickening properties of eggs

Custards: Custard is prepared from a mixture of egg, milk and sugar. It may be cooked over hot water and stirred as it is cooked for soft custard. When custard mixture is cooked without stirring it is baked custard.

Gelation temperature of custards; the temperature at which gelation starts varies with 1. Proportion of ingredients 2. The rate of cooking 3. The quality of eggs

Sugar and egg; increasing the proportion of sugar raises the gelation temperature and gives less stiff custard. Increasing the concentration of egg lowers the gelation temperature and gives stiffer custard.

Rate of cooking: When the custard is cooked slowly, gelation can be perceptible at 78^{0}c and custard attains a thick consistency from 80^{0}C to 84^{0}C. If the cooking is rapid, the custard requires careful watching otherwise curdling will occur.

Quality of eggs; the quality of eggs affects the consistency of custard. Fresh eggs give stiffer whites than aged eggs.

Preparation of soft custard; milk-1 cup-(240g); egg-1(48g); sugar-2 Tb (25g); Mix the ingredients thoroughly till the sugar is dissolved. Heat the mixture in a water bath and stirring it slowly. The heating can be rapid till the custard mixture attains the temperature of 70^{0}C. After this, the heating should be slow so that it takes 3 minutes for 1^{0}C rise in temperature. When the temperature reached 84^{0}C, the custard would have attained the correct consistency.

Baked custard milk-1 cup-(240g); egg-1(48g); sugar-2 Tb (25g); Baked custard is prepared without stirring. Mix the ingredients thoroughly till the sugar is dissolved and the egg has blended well with milk. Keep the mixture in a pan of water and place it in an oven at 180^{0}C. Water bath helps in uniform cooking of the custard. Record the temperature of the custard and continue cooking till it reaches 84^{0}C-85^{0}C and the custard would have attained the correct consistency. This is found out by inserting a knife into the custard and taking it out. The custard should not stick to the knife.

Table 9. Evaluation of thickening properties of egg

Thickening property	doneness of the product well cooked/under cooked	Gel characteristics tender gels/ hard gels	Viscosity Taste uniform flavor/egg flavor	Thick custard / thin custard
Soft custard	....	....	....	....
Baked custard	....	....	....	....
Custard with stored eggs	....	....	....	....

Egg cookery 2

9. Binding properties of eggs

Procedure: Demonstrate the proteins of eggs function as binding material in any two recipe preparation cutlets and croquettes or any other.

Table 10. Evaluation of binding properties of eggs in two recipes

Binding property	Surface binding	Total product binding	Crunchiness	Eating property
Recipe 1	....	....	....	....
Recipe 2	....	....	....	....

10. Coating properties of eggs

Procedure: The material is dipped in beaten egg and then cooked in hot fat. The proteins of egg coagulate and form a thin coating on the surface which prevents penetration of fat effectively.

Table 11. Evaluation of Coating properties of eggs

Coating property	Outcome of the product/ leaching of ingredients/ or not	Crunchiness of product	Color of product	Taste of product
Recipe 1	....	....	....	....
Recipe 2	....	....	....	....

11. Emulsifying properties of eggs

Demonstrate salad dressing 'mayonnaise' recipe to show the emulsifying properties of eggs and mayonnaise is a semisolid emulsion and the egg yolk function as an emulsifying agent.

Basic Procedure

Ingredients: Vegetable oils-55-65 gm, egg yolks-1-2 or whole egg, vinegar-15ml, mustard powder-1tsp, salt-to taste, sugar-1 tsp and other optional spices. The ingredients are combined by agitation by adding oil in a small amount in each time of beating, like wise beat all the ingredients till all oil completes. Mayonnaise should contain not less than 65 per cent oil.

French dressing ingredients required are vinegar, (winterized) oil-35%, salt and spices. The ingredients are combined by agitation. The emulsion is not stable as mayonnaise.

Cooked salad dressing; are a mixture of egg, vinegar, cooked starch, salad oil, and seasonings. Milk may also be added. The starch is cooked and the starch paste is mixed with egg or egg yolk and mixed with oil and seasonings to form an emulsion. It may also contain fruit juices in place of vinegar

Table 12. Evaluation of emulsifying properties of egg

Type of salad dressing	Stable emulsion	Unstable emulsion	Taste/flavor	Color and appearance
Basic				
French dressing	....	....	....	....
Cooked salad dressing	....	....	....	....

12. Meringues

Soft Meringue

Ingredients required are; egg whites- two (60g), sugar powder-4 table spoons, salt-1 g, and vanilla flavor-1 tsp. Beat the egg white until it is foamy. Add salt and vanilla essence and 1 tsp sugar and beat for 2 min. add the remaining sugar in 3 lots and beat for 2 min after each addition. Stiff stable foam is obtained after 15 minutes beating. Bake the meringue at 205^{0}c for 8 minutes.

Hard meringue

Ingredients required are; egg whites- two (60g), sugar powder-100 g, cream of tartar-1 g, vanilla essence-2 ml, salt-1 g. Beat the egg white for 3 minutes. Add the salt, cream of tartar and vanilla essence and 20 g sugar powder and beat for 2 minutes. Add the remaining sugar in 3 lots and beat for 2 minutes after each addition of sugar. After a total beating for 15 minutes stable foam is formed. Then the meringue is baked at a temperature of 135^{0}c for 1 hour.

Table 13. Evaluation of meringues

Type of meringue	Outcome of the product	Stiffness or tenderness of product	Taste/flavor	Color and appearance
Soft Meringue	....	....	....	....
Hard meringue	....	....	....	....

8

Meat Cookery

Procedures for methods of cooking meat and poultry and fish

(Ref: 1997-2016 American Meat Science Association)

The methods used for cooking meat can be divided into two groups:

1. Dry heat methods
2. Moist heat methods

Dry heat methods

Dry heat methods of cooking are suitable for tender cuts of meat or less tender cuts which have been marinated. Dry heat methods include roasting, oven broiling, grilling, pan-broiling, pan-frying and stir-frying.

Roasting: Rub the roast especially the cut edges with one-one mixture of flour and salt and keep the fat side up in a shallow pan and insert a meat thermometer in such a way that the bulb comes near the centre of meat. Fry the meat at 260°C for 15 minutes and then place the meat in another oven at 148.9°C. Continue the cooking until the thermometer reads 61°C and 68°C or 75°C for rare, medium well done meat respectively.

Broiling: Broiling method of cooking is by exposing food to direct radiant heat, either on a grill over live coals or below a gas burner or electric coil. Broiling differs from roasting and baking in that the food is turned during the process so as to cook one side at a time. Temperatures are higher for broiling than for roasting; the broil indicator of a household range is typically set around (288°C), whereas larger commercial appliances broil between (371 and 538°C).

Fish and red meats are suitable for broiling. Steaks, popularly broiled over coals, can also be broiled in skillets or in the oven set on a seasoned wooden plank. In preparation, meats are garnished with vinegar, oil, and minced garlic before being placed on a rack and oven-broiled.

Broiling is suitable for tender meat steaks, ribs and ground meat. Steaks and chops should be at least 3/4 inch thick and ham should be at least 1/2 inch thick for successful broiling. Less tender cuts such as flank steak, top round, and veal, lamb shoulder chops may also be broiled when marinated. Marinating can increase the tenderness of these cuts but only to a limited degree. The same tender cuts suitable for oven broiling can be pan- or griddle broiled. This method is especially good for meat 3/4 inch or less in thickness; very thick cuts of meat may become overcooked on the outside before the middle has reached the desired degree of doneness.

Pan-broiling: In this method the meat is cooked in the pan in the oven. Pan-broiling is a faster and more convenient method than oven broiling for cooking thinner steaks or chops.

Grilling (Barbecuing): The technique we call grilling is thought to have originated in the Caribbean, where natives smoke-dried meat over hot coals on wood-frame grills. Early Spanish explorers called this the "barbacoa" which evolved into the modern-day word "barbecue."

Meat can be grilled on a grid or rack over coals, heated ceramic briquettes or an open re. While it is usually done out- doors, grilling can be done in the kitchen with special types of range tops or newer, small appliances.

Standard charcoal briquettes are the most common fuel for grilling. High-quality briquettes burn evenly and consistently. Flammable material for quick-start fires may be added. It takes longer for natural lump charcoal to get hot, but it provides heat for a longer period of time.

Wood chips are first soaked in water about 30 minutes, drained, and then placed on the burning coals. (Softwoods and evergreens should not be used; they can impart a bitter avor and leave a residue in the grill.)

Grilling is often used to cook kabobs. Kabobs are pieces of meat, or a combination of meat and vegetables, or meat and fruit pieces, alternated on a skewer.

Stir-frying: Stir-frying is taking small or thin pieces of meat and stirred almost continuously in the pan and cooking is done with high heat.

Deep-fat frying: When meat is cooked immersed in fat, the process is called deep-fat frying. This method is only used with very tender meat. Usually, meat to be deep-fat fried is coated with egg and crumbs or a batter, or it is dredged in flour or corn meal (breaded). This method of cooking is sometimes used for sweetbreads, liver and croquettes; however, a number of other meat products are suitable for deep-fat frying.

Pan-frying: Pan-frying is that a small amount of fat is added first, or allowed to accumulate during cooking. Pan-frying is a method suitable for ground meat, small or thin cuts of meat, thin strips, and pounded, scored or other- wise tenderized cuts that do not require prolonged heating for tenderization.

Moist Heat methods

Moist-heat methods of cooking are suitable for less tender cuts of meat. Moist-heat cooking helps to reduce surface drying in those cuts requiring prolonged cooking times. Unless a pressure cooker is used, cooking temperature is usually low, but heat penetration is faster than in dry-heat methods because steam and water conduct heat rapidly.

With moist-heat cookery, meat may lose some water-soluble nutrients into the cooking liquid. However, if the cooking liquids are consumed, as in stews or soups, nutrients are transferred and not totally lost.

Braising-Meat can be braised in cooking bags designed specifically for use in the oven. Use of oven-cooking bags can reduce cooking time for larger cuts of meat. No additional water is needed, as moisture is drawn out of the meat due to the atmosphere created by the cooking bag.

Cooking in Liquid-Less tender cuts of meat can be covered with liquid and gently simmered until tender. Care should be taken not to let the temperature of the liquid exceed 90.6^{0}c because boiling (100^{0}c) toughens meat protein. When the liquid is used as a base for soup it is called meat stock (also called broth or bouillon). Meat that is partially cooked in liquid before cooking by another method is called "parboiled."

The three ways to cook in liquid are **simmering, stewing and poaching**. Simmering and stewing are used for less tender cuts of meat while poaching is used for tender cuts. The difference between simmering and stewing is that **simmering** is used with whole cuts of meat while stewing is used with small pieces of meat.

Poaching has been a traditional way of cooking poultry and fish. However, meat roasts can also be successfully poached if they come from tender cuts.

After an initial browning period, the poaching liquid is added and the roast is then gently simmered until it reaches 54.4^{0}c. A combination of meat broth, red wine and herbs makes a flavorful poaching liquid. After cooking, the liquid can be used to make a simple sauce for the roast or it can be strained and frozen for later use as a soup base or stewing liquid.

Poaching takes one third less time than roasting. (A meat roast will poach to rare in about 20 to 30 minutes). In addition to cooking more quickly, poaching helps to keep shrinkage of the meat to a minimum A poached roast is also just as tender, juicy and avorful as one which has been conventionally prepared.

Pressure cooking; Meat can be cooked in a steam in a pressure cooker

Cooking meat and poultry and fish

There are two basic methods of cooking meat:

- Moist heat
- Dry heat

It is important to select the proper cooking method for the cut of meat. Less tender cuts of meat require moist heat cooking methods to help break down the tough connective tissues. Moist heat cooking means moisture is added to the meat and the meat is cooked slowly over a long time; it includes:

- Braising, and
- Cooking in liquid, such as stews or other slow cooker recipes.

Tender cuts of meat do not require moisture and long, slow cooking. They are usually cooked with a dry heat method, including:

- Roasting,
- Broiling,
- Pan-broiling,
- Pan-frying, and
- Grilling.

The method chosen to cook a certain cut of meat should relate directly to the inherent tenderness of that cut. Tenderness is determined by:

- Where on the animal the meat comes from,
- The degree of marbling,
- The age of the animal,
- How the meat was stored, and
- How the meat was prepared for market.

In general, cuts from the loin section are the tenderest; the farther away from this section the less tender the meat will be.

Cooking Tender Cuts of Meat

Roasting

Roasting is a cooking method in which meat is surrounded and cooked by heated air, usually in an oven. Meat is not covered and no water is added. Follow these steps:

- Place meat fat side up on a rack in a shallow open roasting pan.

- Season as desired.
- Insert meat thermometer; be sure tip does not rest in fat or on a bone.
- Do not add water. Do not cover.
- Roast in a slow oven at 325°F until the thermometer reaches the desired doneness.
- Baste with drippings during cooking.

To test for doneness, use a meat thermometer. The internal temperature shows exactly how done the meat is. Look up the roasting time tables in a cookbook. The more tender cuts of meat will remain tender if cooked to rare rather than well-done. On the other hand, less tender cuts may be tenderer if they are cooked to medium or well-done, rather than rare.

Broiling, Pan-broiling, or Pan-frying

The basic rule for broiling, pan-broiling or pan-frying meat is to use enough heat to brown the outside without overcooking the inside of the meat. A moderate temperature is best for broiling and frying most meats.

Broiling

Broiling is cooking by direct heat from a flame, electric unit, or glowing coals. Meat is cooked one side at a time. Choose tender beef steaks, lamb chops, cured ham slices, and bacon for broiling. Use steaks or chops cut 1 to 2 inches thick. If steaks or chops are less than 1 inch thick pan-broil them.

Consult the manufacturer's instructions for broiling since equipment varies. Usually the door is left open when broiling in an electric range and closed when broiling in a gas range.

- Place meat on a rack in a broiler pan.
- Place pan two to five inches from heat. The thicker the cut, the farther the meat should be placed from the heating unit to assure even cooking.
- Broil one side until browned. Season cooked side, if desired.
- Turn meat; cook second side to desired doneness and until meat is browned. Season, if desired.

Pan-broiling

Pan-broiling is cooking in an uncovered pan over direct heat. Fat that cooks out of the meat is drained off.

- Place meat in preheated heavy frying pan.

- Do not add oil or water. Do not cover.
- Cook slowly, turn occasionally. Pour off fat as it accumulates.
- Cook to desired doneness, until both sides are browned.
- Season, if desired.

Pan-frying

Pan-frying is similar to pan-broiling, except that meat is cooked in a small amount of fat.

- Heat a small amount of oil in a skillet over medium heat.
- When oil is hot, add meat; do not cover.
- Turn occasionally until done as desired and browned on both sides.
- Season, if desired.

The easiest way to tell when steaks and small pieces of meat are done when you broil, pan-broil, or panfry is to make a small cut in the meat near the bone and check the interior color.

- Rare beef will be reddish pink with lots of clear red juice.
- Medium beef has a light pink color and less juice than rare.
- Well-done beef is light brown with slightly yellow juice. Fresh pork should be cooked until the juice is no longer pink.

Cooking Less Tender Cuts

Braising

Braising is cooking in steam trapped and held in a covered container or foil wrap. The source of the steam may be water or other liquid added to the meat, or it may be meat juices. Large, less tender cuts, such as chuck, round, and rump, are braised as pot roasts.

- In a heavy frying pan, brown meat on all sides in a small amount of oil; pour off fat.
- Season, if desired.
- Add a small amount of liquid to the meat; cover pan tightly.
- Simmer on top of the range or cook in the oven at 350°F until tender.

Cooking in Liquid

this method involves covering a less tender cut of meat with liquid and simmering in a covered kettle until tender and well-done.

- In a Dutch oven or heavy pan, brown meat on all sides in a small amount of oil; pour off fat.
- Season, if desired.
- Add enough liquid to cover meat completely; cover pan tightly.
- Simmer on top of the range or in the oven until tender.
- Add vegetables just long enough before serving to be cooked.

Cooking Poultry

The type of method to use for cooking poultry depends on the bird. Young poultry is best for roasting, broiling, and frying. Older poultry requires braising or stewing methods. Either way, slow, even heat should be used for tender, juicy, evenly done poultry. Do not overcook; it results in tough, dry meat.

Broiling

- Cut chicken broiler in half lengthwise, in quarters, or in pieces. Quarter young turkey fryers or roasters, or cut in pieces.
- Fold wing tips across back side of poultry quarters.
- Set oven control to broil.
- Brush poultry with one tablespoon margarine or butter.
- Place poultry skin side down on rack in broiler pan.
- Place broiler pan so top of chicken is seven to nine inches from heat.
- Broil 30 minutes. Sprinkle with salt and pepper.
- Turn chicken and brush with one tablespoon margarine or butter. Broil 15 to 25 minutes longer or until chicken is brown and juices run clear.

Poaching (in the microwave)

An easy way to be prepared for any recipe that calls for cooked chicken is to poach chicken in the microwave ahead of time and have it stored in the freezer. That way, cooked chicken is available for use in casseroles, sandwiches, and salads.

- Place four chicken breast halves, skin side up, in a 12x8-inch (2 quart) microwave-safe baking dish, with the thickest portions placed toward the outside edges of the dish. If desired, sprinkle the chicken lightly with seasoned salt, paprika, and pepper.
- Cover the dish with waxed paper. Microwave on HIGH for 12 to 14 minutes or until it is fork tender and juices run clear.

- Use the chicken immediately, or cool it completely before removing the meat from the bones. Package the cooked chicken in freezer bags or containers in the amounts needed in recipes. Store in the refrigerator up to two days or in the freezer up to two months.
- Thaw the frozen cooked chicken in one of two ways:
    - o Place chicken in a microwave-safe covered casserole and microwave it on DEFROST for four to six minutes or until the chicken is thawed. Break up and rearrange the chicken halfway through thawing. When thawed, the chicken will feel cool to the touch.
    - o Leave chicken in its moisture/vapor resistant freezer container and thaw overnight in the refrigerator.

      Note: Chicken breasts can also be poached in a large saucepan on top of the stove. Add cold water just to cover chicken, bring to a boil, and simmer for 30 to 40 minutes or until chicken is tender. Skim off any scum that rises to the surface.

Roasting

- Place poultry breast side up on a rack in a shallow roasting pan. Do not add water. If desired, brush poultry with cooking oil or melted margarine or butter.
- Cover poultry with a loose tent of heavy-duty aluminum foil. To make a tent, tear off a sheet of foil 5 to 10 inches longer than the poultry. Crease foil crosswise through the center and place over the bird, crimping loosely onto sides of pan to hold it in place. This prevents overbrowning, keeps the bird moist, and reduces oven spatter.
- Insert a meat thermometer through the foil into the thickest part of the thigh muscle without touching the bone. The inner thigh is the area that heats most slowly. For turkey parts, insert the thermometer in the thickest area.
- Roast at 325°F according to the timetable. To brown poultry bird, remove the foil tent 20 to 30 minutes before roasting is finished, and continue cooking until the thermometer registers 185°F.
- Basting is usually not necessary during roasting since it cannot penetrate the turkey; it does help brown the skin.

Oven Cooking Bags

Preparing poultry in an oven cooking bag is a moist heat cooking method. It is the best way to produce a moist, tender bird. It also helps reduce oven spatter. Using ordinary brown bags for roasting is not recommended because they may not be sanitary, the glue and ink used on brown bags have not been approved for use as cooking materials, and the juices formed as the poultry cooks may saturate the bag and cause it to break.

- Preheat oven to 350°F.
- Shake 1 tablespoon of flour in the bag to prevent bursting.
- Place celery and onion slices in the bottom of the bag to help prevent poultry from sticking to the bag and to add flavor.
- Place poultry on top of the vegetables, close the bag with the enclosed twist-tie, and make 6 half-inch slits in the top to let steam escape. Insert meat thermometer through a slit in the bag. The poultry is done when it reaches 180°F.
- When poultry is done, cut or slit the top of the bag down the center. Loosen the bag from the turkey so there is no sticking and carefully remove the poultry to the serving platter.

  For a printer-friendly version of this chart, click on the icon to download. **Note:** Must have Acrobat Reader.

Fish Cooking Methods

The three secrets to success fish cooking are;

1. Do not overcook fish as the texture will coarsen, dry out and the flavor will be destroyed.
2. Don't over spice fish as it has very delicate flavors so be light-handed with herbs and spices.
3. Keep fish moist and preserve the natural juices whilst cooking by using a moist cooking method or baste frequently during dry cooking.

Cooking Methods

Baking

Baking in a moderate oven 180-200°C is an extremely useful method of cooking whole fish, fillets, cutlets or steaks. But remember it is a dry heat method and fish, especially without its skin, tends to dry out, so use baste, marinade or sauce to reduce the moisture loss.

Baking in Foil

Baking in foil is an excellent way to retain flavor and moisture particularly of larger steaks, cutlets or whole fish. Use a liquid such as fish stock, white wine or lemon juice with a little butter, salt, pepper and seasonings of your choice before sealing the fish in foil. Bake in a moderate oven 180-200^{0}C for a mouth-watering result

Barbecuing

Barbecue fish, but protect it with marinades, bastes, lemon juice or oil or butter brushed on frequently during the cooking. Or wrap it in foil with these liquids and seasonings. Be careful turning the fish over during cooking.

Casseroling

Casseroling in a little liquid in a covered dish in a moderate oven will achieve a dish in which the subtle differences in the flavors of the various species and their natural juices are maintained.

Shallow Frying

Shallow frying is cooking in a small quantity of fat sufficient to come up to the level of half the thickness of the fish - in a wide shallow pan. The best fat for fish is butter or half butter and half olive oil. The oil combined with the butter reduces the risk of overheating the butter.

Deep Frying

Deep frying is immersing the fillet or whole fish in deep oil in a deep pan after protecting it first with a coating such as egg and breadcrumbs or batter. The oil must be at the correct temperature (175-195C) before placing the fish into it. If the oil isn't hot enough the coating will soak up the oil and become greasy and if it is too hot the coating will burn before the fish is cooked. To test the temperature without a thermometer the oil should be heated until a faint haze rises from it before dropping a small cube of bread into it. If it rises, bubbling to the surface and becomes golden brown it is ready. If it turns dark brown rapidly the heat must be reduced and if it sinks and stays low in the oil it is not hot enough yet.

Oven Frying

The result is similar to shallow frying but is done in hot butter in a preheated baking dish in a hot oven 230-250C. Because of the high temperature used the fish cooks quickly so this method is best suited to thin whole fish.

Grilling

Grilling is a fast way to cook fish. Using either fillets or whole fish this simple method allows the fish to develop its own rich flavor under the intense heat. Fish should be basted during cooking, either with butter or oil or a prepared baste, to prevent it drying out. Alternatively the fish may be marinated beforehand and the liquid used for basting. Whole fish or thicker fillets seem to fare better under the grill as the fish has time to develop a rich golden brown by the time the inside is cooked. The high heat penetrates and cooks thin uncoated pieces too fast for browning to occur. If whole fish are to be grilled score the skin and flesh to allow better heat penetration.

Marinating

Marinating has two distinct purposes: the first is to impart a flavour by presoaking the fish in a mixture of lemon juice, oils and flavorings and then using the marinade liquid to baste during grilling, barbecuing or baking. The second is to replace the cooking process altogether. Very fresh fish is cut into boneless bite sized pieces and left to soak in lemon juice and other flavorings until the flesh becomes opaque and white right through - approximately 6 to 12 hours in the refrigerator or 4 to 8 hours at room temperature. The marinade is then strained off and the fish added to crisp, finely cut vegetables with dressings or sauces such as mayonnaise, yogurt or coconut cream. It is then served chilled.

Poaching

Poaching as far as the fish is concerned means totally immersing it in seasoned stock or court bouillon. The liquid should be brought rapidly to the point where the surface begins to swirl but with no bubbles rising to surface (boiling should not occur). At this point the heat should be reduced (and sometimes even turned off). Inspect to see if the flesh flakes easily and comes away from the bone. Poaching is unmatched as versatile method of cooking fillets, steaks, cutlets or whole fish of any fleshy type. It can then be served steaming hot, dressed in a sauce made from the poaching liquid or chilled and served in a variety of salads or as a cold entree. It's the perfect cooking method for weightwatchers.

Sousing

Sousing is gently cooking small fish or fillets in a combination of vinegar, water and a selection of various herbs and spices. Although it may be served either hot or cold, the delightful delicate piquancy of the subtle blend of flavors only becomes evident after chilling. Fish cooked this way will keep for up to a week in the refrigerator.

Steaming

There are two types of steaming. In one the fish is put into the upper part of a double saucepan with a perforated base and a tightly fitted lid. Steam rising from boiling water in the saucepan below and passes through the perforations and surrounds the fish cooking in it. In the second type of steaming no steam reaches the fish as the upper part of the saucepan has no perforations (two plates on top of the saucepan can be used if no double saucepan is available.) This method is slower than the first but effective and retains all the natural juices.

Microwave Cooking

Seafood adapts well to microwave cooking - natural flavors and colors are retained and often enhanced by this quick method of cooking. The seafood can be cooked without any additional liquid other than perhaps a small amount of melted butter, lemon juice or wine. Covering fish with tomatoes, lemon or orange slices or herbs will help to retain moisture and enhance the flavor. Grilling and frying fish is possible if a browning dish is used.

Aim: To study the factors affecting color and texture of meat and to learn various methods of cooking of meat.

Procedure: Tender meat-50 g less tender meat-50 g

Boil the meat according to the initial cooking 60^{0}C -65^{0}C, medium-done cooking 65-70^{0}C and well-done cooking 70-80^{0}C.

Record the characteristics of raw and cooked meat color and textural qualities according to the following tables 1 and 2 and record the observations.

Table 1. Effect of heat on meat pigments and color during cooking

Type of meat	**Raw meat color (outer layer-inner layer)**	**Color of (Rare/ initial cooking below 60-65^{0}C) (outer layer-inner layer color)**	**Color of (medium-done cooking 65-70^{0}C) (outer layer-inner layer color)**	**Color of (well-done cooking 70-80^{0}C) (outer layer-inner layer color)**	**Inferences**
Tender meat	....	....	....	....	
Less tender meat	....	....	....	....	

Table 2. Effect of cooking on water holding capacity of meat

Type of meat	**Weight of raw meat (100 g)**	**Weight of (Rare/initial cooking below 60-65^{0}C)**	**Weight of (medium-done cooking 65-70^{0}C)**	**Weight of final (well-done cooking 70-80^{0}C)**	**Inferences**
Tender meat	....	....	....	....	
Less tender meat	....	....	....	....	

Table 3. Effect of Heat on Tenderness of meat proteins during cooking

Procedure: Record the internal and external temperatures of meat pieces

Type of meat	Temperature (^{0}C) of (Rare/ initial cooking below 60-65^{0}C)		Temperature (^{0}C) of (medium-done cooking 65-70^{0}C) 70-80^{0}c)		Temperature (^{0}C) of (well-done cooking		Inferences
	External temp^{0}c	Internal temp^{0}C	External temp^{0}C	Internal temp^{0}V	External temp^{0}V	Internal temp^{0}V	
Tender	....	....	....	....	....	....	
Meat Less	....	....	....	....	....	....	
Tender meat	....	....	....	....	....	....	

Table 4. Effect of heat on flavor of meat proteins during cooking

Type of meat	Flavor of raw meat (100 g)	Flavor of (Rare/initial cooking below 60-65^{0}C)	Flavor of further (medium-done cooking 65-70^{0}C)	Flavor of final (well-done cooking 70-80^{0}C)	Inferences
Tender meat	....	....	....	....	....
	....	....	....	....	....
Less tender	....	....	....	....	....
	....	....	....	....	....
Meat	....	....	....	....	....

Methods of cooking meat

Procedure: Tender meat-50 g less tender meat-50 g

Dry methods: Boil the meat according to the methods given in table and record the cooking characteristics of cooked meat

Table 5. Dry methods of cooking and their effects on meat texture

Dry-media method	Changes in Texture in less tender meat	Changes in Texture in tender meat	Changes in color	Cooking Temperature (^{0}C)	Cooking Time (minutes)	Infer-ences
Roasting	....	....	....	....	....	....
	....	....	....	....	....	....

Broiling	....	....	....	....	....	....
	....	....	....	....	....	....
Pan-broiling	....	....	....	....	....	....
	....	....	....	....	....	....
Pan Frying	....	....	....	....	....	....
	....	....	....	....	....	....
Deep Fat Frying	....	....	....	....	....	....

Moist methods

Procedure: Tender meat-50 g less tender meat-50 g

Moist methods: Boil the meat according to the methods given in table and record the cooking characteristics of cooked meat.

Table 6. Moist methods of cooking and their effects on Meat Texture

Moist-Media methods	Changes in Texture (Rare/initial cooking below 60-65^{0}C)	Changes in Texture (medium-done cooking 65-70^{0}C)	Changes in texture (well-done cooking 70-80^{0}C)	Final Cooking Temper-ature (^{0}C)	Final Cooking Time (minutes)	Infere-nces on color, flavor and taste
Braising in sauce pan	....	....	....	....	....	....
Braising in pressure sauce pan	....	....	....	....	....	....
Braising in Foil Wrap	....	....	....	....	....	....
Braising in plastic Bag	....	....	....	....	....	....
Stewing meat	....	....	....	....	....	....
Cooking in Water	....	....	....	....	....	....

Cooking in	....	....	....	....	....	....
Steam	....	....	....	....	....	....
Cooking	....	....	....	....	....	....
meat in	....	....	....	....	....	....
simmering	....	....	....	....	....	....
for 3-4 hrs	....	....	....	....	....	....
Pressure	....	....	....	....	....	....
cooking	....	....	....	....	....	....
Cooking	....	....	....	....	....	....
Frozen	....	....	....	....	....	....
meat	....	....	....	....	....	....

Tenderness of meat and factors affecting it

Procedure: Tender meat-50 g less tender meat-50 g

Boil the meat according to the tenderization methods given in the table and record the cooking characteristics of cooked meat.

Table 7. Factors Effecting Tenderness of meat

Variation in tenderness/ tenderizing agents	Cooking time (minutes)	Textural changes	Color changes	Flavor changes	Inferences
Ageing of meat store the	....	....	....	....	....
Meat at 0^0c for 48 hours	....	....	....	....	....
Mincing/grinding/	....	....	....	....	....
pounding Meat	....	....	....	....	....
	....	....	....	....	....
Adding proteolytic enzymes	....	....	....	....	....
(papain) or cooking Meat	....	....	....	....	....
with raw papaya piece	....	....	....	....	....
	....	....	....	....	....
Wrapping Meat in papaya	....	....	....	....	....
Leaves for 1 hour	....	....	....	....	....
	....	....	....	....	....
Freezing of Meat	....	....	....	....	....
for 24 hours	....	....	....	....	....

Effect of Acid	....	....	....	....	....
Soak the meat in sour curd for 1 hour	....	....	....	....	....
Soak the meat in sour tomato juice for 1 hour	....	....	....	....	....
Soak the meat in Vinegar for 1 hour	....	....	....	....	....
Cooking meat at **High temperature (pressure cooking)**	....	....	....	....	....
Effect of mineral salts	....	....	....	....	....
Soak the meat in salt solution for 72 hours in refrigerator	....	....	....	....	....
Soak the meat in a mixture of mono and di-hydrogen sodium phosphates with a pH of 6.0	....	....	....	....	....
Soak the meat in dilute calcium chloride solution	....	....	....	....	....
Change in meat pH to 4.0 by appropriate solution	....	....	....	....	....

9

Poultry Cookery

Aim: To study and to learn various methods of cooking of poultry

Procedure: Poultry meat-50 g for each method

Cook the poultry meat according to the methods given in the table and record the cooking effects

Table 1. Methods of cooking poultry and factors affecting it

Methods of cooking	Cooking time (minutes)	Tenderness of meat	Juiciness of meat	Uniformity in cooking	Color/ flavor
Dry- Heat methods					
Roasting					
Broiling					
Frying					
Cooking breast-side down					
Basting					
Stuffing					
Grilling					
Moist-Heat methods					
Braising					
Stewing					

Cooking in water

Simmering

Grilling

Any other

10

Fish Cookery

Aim to learn different methods of cooking Fish and factors affecting it

Procedure: Takes clean, washed and edible fish-50-100 g for each method of cooking. And cook the fish according to the methods given table and record the characteristics of cooked fish

Table 1. Methods of cooking Fish and factors affecting it

Methods of cooking	Under cooked/over cooked/well done	Acceptable golden brown color	Flavor/Taste Pleasant flavors/No flavors	Mouth watering feeling	Soft texture/coarse texture/dry texture
Baking	....	....	....	....	....
	....	....	....	....	....
Broiling	....	....	....	....	....
	....	....	....	....	....
Cooking in water	....	....	....	....	....
	....	....	....	....	....
Cooking in	....	....	....	....	....
	....	....	....	....	....
Steam	....	....	....	....	....
	....	....	....	....	....
Pan-Frying	....	....	....	....	....
	....	....	....	....	....
Deep-Fat frying	....	....	....	....	....
	....	....	....	....	....
Poaching	....	....	....	....	....
	....	....	....	....	....
Microwave	....	....	....	....	....

	....	....	....	....	....
Cooking	....	....	....	....	....
	....	....	....	....	....
Sousing	....	....	....	....	....
	....	....	....	....	....
Marinating and	....	....	....	....	....
cooking	....	....	....	....	....

11

Fats and Oils Cookery

Aim to study the smoking point of different oils, oil absorption factors in deep fat frying and shortening effect of fat in food preparation

1. Properties of fats and oils

Smoking Point of some Fats and Oils

Procedure: Measurement of smoke point of given oils: Heat the given oils in a deep frying pan till fumes come out from oil. Record the temperature when the fumes come out and noted it as a 'smoke point'. Cool the oil by keeping oil pan at room temperature. After cooling repeat the heating of same oils 2nd and 3rd time to measure the smoke point and record the observations

Table 1. Smoking Points of some Fats and Oils

Name of Fat or Oil	Smoking temperature ^{0}C during		
	First heating of oil/fat	Second heating of oil/fat	Third heating of oil/fat
Butter Fat	....	....	....
Ghee	....	....	....
Vanaspathi	....	....	....
Coconut oil	....	....	....
Peanut oil	....	....	....
Sunflower oil	....	....	....
Gingelly oil	....	....	....

Soyabean oil	....	....	....
Rice bran oil	....	....	....
Olive oil	....	....	....

2. Factors affecting oil absorption during deep fat frying

1. Effect of Temperature and Time of cooking on oil absorption

Procedure: prepare recipes and deep fry the product at three different temperatures 170°c, 185°c and 200°c and note down the oil absorption characteristics and record the observations

Table 2. Effect of Temperature and Time of cooking on oil absorption

Type of food for frying	Weight of oil before frying food (gms) A	Temperature of frying °C	Time taken for cooking (minutes)	Weight of oil After frying food (gms)B	Total oil absorbed by food (gms) (B-A)
Doughnuts/Muruku/ Meat piece		170°C			
		185°C			
		200°C			

2. Effect of total Surface area of food frying on oil absorption

Procedure: prepare less surface area, moderate surface area and high surface area food recipes and deep fry the products and record the oil absorption characteristics

Table 3. B. Effect of total Surface area of food frying on oil absorption

Type of food for frying	Surface area of food	Weight of oil before frying food (gms) A	Temperature of frying ^{0}C	Time taken for cooking (minutes)	Weight of oil After frying food (gms) B	Total oil absorbed by food (gms) (B-A)
Chegodi/ Gavvalu/	Less surface area	....	....	....	....	....
Biscuits/pappu billalu	Moderate surface area	....	....	....	....	....
Ariselu/ Nippatlu/ Poli	Large surface area	....	....	....	....	....

3. Effect of moisture content of the dough on oil absorption

Procedure: prepare low moisture, moderate moisture and high moisture food recipes and deep fry the products and record the oil absorption characteristics

Table 4. C. Effect of Moisture content of the dough on oil absorption

Type of food for frying	Moisture content of food	Weight of oil before frying food (gms) A	Temperature of frying ^{0}C	Time taken for cooking (minutes)	Weight of oil After frying food (gms) B	Total oil absorbed by food (gms) (B-A)
Pappu billalu	Less moisture	....	....	....	....	....
Poori/vada/ garelu	Moderate moisture	....	....	....	....	....
Bajji/burelu/ bonda	High moisture	....	....	....	....	....

4. Effect of Protein Fat and carbohydrate content of food on oil absorption

Procedure: plan a recipe which is high and low in protein or high and low in fat or high and low in carbohydrate and deep fry the product and record the practical observations as given table

Table 5. D. Factors affecting Protein Fat and carbohydrate content of food on oil absorption

Type of food for frying	Protein Fat and carbohydrate content of food	Weight of oil before frying food (gms) A	Temperature of frying ^{0}C	Time taken for cooking (minutes)	Weight of oil After frying food (gms) B	Total oil absorbed by food (gms) (B-A)
	High protein food	....	....	....	....	....
	Low protein food	....	....	....	....	....
	High Fat food	....	....	....	....	....
	Low Fat food	....	....	....	....	....
	High carbohydrate food	....	....	....	....	....
	Low carbohydrate food	....	....	....	....	....

5. Effect of low, moderate and high smoke point of oils on oil absorption in poori preparation

Procedure: prepare poori dough's and deep fry it using high, moderate and low smoke point oils and record the oil absorbed by poori in a given table

Table 6. Effect of low, moderate and high smoke point of oils on oil absorption in poori preparation

Type of food for frying	Smoking temperature of fat/oil	Weight of oil before frying food (gms) A	Temperature of frying ^{0}c	Time taken for cooking (minutes)	Weight of oil After frying food (gms) B	Total oil absorbed by food (gms) (B-A)
Poori	Low Smoking temperature of fat/oil	....	....	....	....	....
		....	....	....	....	....
		....	....	....	....	....
		....	....	....	....	....
		....	....	....	....	....
Poori	Moderate Smoking temperature of fat/oil	....	....	....	....	....
		....	....	....	....	....
		....	....	....	....	....
		....	....	....	....	....
		....	....	....	....	....
Poori	High Smoking temperature of fat /oil	....	....	....	....	....
		....	....	....	....	....
		....	....	....	....	....

3. Preparation of Whipped Cream as double emulsion

Principle: Whipped cream is air-in-water foam in which air cells are surrounded by a film containing fat droplets stabilized by a film of protein. Partial denaturation of this protein occurs as the cream is whipped. There is some clumping of the fat globules in the cell walls of the foam, and the fat is partially solidified, preventing collapse of the cell walls.

Most whipped creams also contain a stabilizer such as carrageenan or sodium alginate to increase the viscosity of the aqueous phase, which retards any tendency to creaming or syneresis.

Definition of whipped cream

Whipped cream is **liquid heavy cream** that is whipped by a whisk or mixer until it is light and fluffy and holds its shape, or by the expansion of dissolved gas, forming a firm colloid. It is often sweetened, typically with white sugar, and sometimes flavored with vanilla.

Examples of emulsion

Familiar foods illustrate examples: **milk is oil in water emulsion**; **margarine** is water in oil emulsion; **and ice cream** is an oil and air in water emulsion with solid ice particles as well. Other food emulsions include mayonnaise, salad dressings, and sauces such as Béarnaise and Hollandaise.

Properties of whipped cream

Whipped cream is cream or heavy cream that has been whipped until the texture changes from a fatty liquid to a light and fluffy foam. As the cream is whipped, air bubbles are incorporated into the fat, resulting in an airy mixture that is approximately **double the** volume of the original liquid.

Solution: Whipped cream from milk is an example of **colloidal solution** in which the dispersed phase is gas and the dispersion medium is liquid.

Whipping of Milk Products

Evaporated milk can be successfully whipped if it is thoroughly chilled prior to whipping. Air bubbles incorporated during whipping are trapped in the viscous liquid to form foam. The difference in initial viscosity partially accounts for the difference in stability of foams formed from milk and from the more concentrated evaporated milk.

Nonfat dry milk can be whipped if reconstituted to a higher than normal solids content.

Table 7. Evaluation of whipping of milk products

Type of milk product	Formation of whip/solution/ colloidal/foam	Oil-in-water emulsion	Water-in-oil emulsion	Appearance – increase in volume/ no increase
Evaporated milk	....	....	....	....
	....	....	....	....
Non-fat-dry milk	....	....	....	....
	....	....	....	....
Standard milk	....	....	....	....
	....	....	....	....
Any Market	....	....	....	....
samples/ice	....	....	....	....
cream/ margarine	....	....	....	....

4. Preparation of basic cakes with different types of fats

1. Preparation of basic cake with Butter Fat
2. Preparation of basic cake with Hydrogenated Fat
3. Preparation of basic cake with Margarine

Table 8. Evaluation of Cakes prepared with different types of fats

Type of cake	Cake setting property	Crumb structure	Crust structure	Taste and flavor
Basic cake with Butter Fat	....	....	....	....
	....	....	....	....
	....	....	....	....
Basic cake with	....	....	....	....
Hydrogenated Fat	....	....	....	....
	....	....	....	....
Basic cake with Margarine	....	....	....	....

5. Preparation of salad dressings with olive oil and other vegetable oils

1. **Lemon Balsamic**: Whisk 2 tablespoons balsamic vinegar, 1 tablespoon lemon juice, 2 teaspoons mustard, 1/2 teaspoon kosher salt, and pepper to taste. Gradually whisk in **1/2 cup olive oil.**
2. **Hazelnut-Herb:** Blend 2 tablespoons each mustard and cider vinegar, 1 teaspoon kosher salt, and 1/3 cup each **vegetable oil and hazelnut oil** in a blender. Add 1/4 cup each chopped chives and dill and pulse to combine.

3. **Lemon: Whisk** 2 tablespoons lemon juice, 1 tablespoon dijon mustard, 1 teaspoon lemon zest, 1/2 teaspoon sugar, and salt to taste. Gradually whisk in 1/4 cup each O**live oil and vegetable oil**
4. **Spiced Chutney:** Whisk 2 table spoons mango chutney and lime juice and 1/2 teaspoon each ground cumin and kosher salt. Gradually whisk in **1/4 cup vegetable oil.**
5. **Mango-Lime**: Purée 1 chopped peeled mango, the zest and juice of 1 lime, and 1 teaspoon each mustard, sugar and kosher salt in a blender. Gradually blend in 1/4 cup rice vinegar and **1/2 cup vegetable oil.**

Table 9. Evaluation of salad dressings

Type of salad dressing	Stability of emulsion	Flavor	Sourness	Oiliness
1.	....	....	....	....
2.	....	....	....	....
3.	....	....	....	....
4.	....	....	....	....
5.	....	....	....	....

12

Sugar Cookery

Sugar cookery-1

Properties of cane sugar

Aim: To learn and study the Properties of cane sugar, crystallization of sugar and stages of sugar cookery

Principle: Pure sugar when heated melts at about 160-180^{0}c. at about 210^{0}c sucrose loses water and becomes caramel

1. Factors affecting Solubility of sucrose in water at different temperatures

Procedure: Dissolve sucrose in 100 g of water as per the figures given in table 1 and record the practical observations

Table 1. Solubility of sucrose in water at different temperatures

Temperature ^{0}C	Sucrose (g) per 100 g of water (Book values)	Find out the laboratory value solubility of sucrose (g) per 100 g of water/respective temperature
0	179.2	
10	190.5	
20	203.9	
30	219.5	
40	238.1	
50	260.4	
60	287.3	

70	320.5
80	362.1
90	415.7
100	487.2

2. Boiling point of sugar solution

Procedure: When a substance is dissolved in water the boiling point of the solution increases depending upon the quantity of substance present in it. Find out the boiling point of sucrose solution as per figures given in table 2 and record the observations

Table 2. Boiling point of sugar solution

Sucrose (g) per 100 g syrup	Boiling point ^{0}C (Book values)	Find out the laboratory value Sucrose (g) per 100 g syrup/respective boiling point
10	100.4	
20	100.6	
30	101	
40	101.5	
50	102	
60	103	
70	106.5	
80	112	
90	130	

4. **Saturated sugar solution:** One hundred gram of water can dissolve 487.2 g of sugar at 100°C and one hundred gram of water can dissolve 260.4 g at 50°C to form saturated solution of sugar at 100°C and 50°C respectively.
5. **Super saturated sugar solution:** If the saturated solution of sugar at 100°C containing (487.2 g of sugar) is cooled to 50°C, at this temperature solution contains excess sugar does not separate immediately and will be held in solution even though water can dissolve 260.4 g of sugar at 50°C. Such solution containing more or excess sugar than that required is called super-saturated solution.

Table 3. Preparation of saturated sugar solution and Super saturated sugar solution

Type of sugar solution	g of water	(Book values)		Find out the laboratory value of Saturated sugar solution
Saturated sugar solution	100 g of water	487.2 g of sugar	100°C	
	100 g of water	260.4 g of sugar	50°C	
				Find out the laboratory value of super saturated sugar solution
Super saturated sugar solution	100 g of water	487.2 g of sugar	100°C But when cooled to	
	-do-	-do- (excess sugar) is present and form super saturated solution	50°C	

Sugar cookery-2

Factors affecting Crystallization of sugar

Principle: Crystallization of sugar occurs from a super saturated solution. Ex; preparation of fondants and fudges are done from super saturated solutions.

Factors affecting the Crystallization of sugar

Formation of nuclei: Nuclei can only form from a super saturated solutions. Nuclei form simultaneously in various places in the solution and crystallization begins from these nuclei. When more number of nuclei are formed in the solution, crystals become more and grow to a small size, when few number of nuclei are formed in the solution, the crystals grow to a large size.

Seeding: When crystals of same material are added to the solution to start crystallization, the process is called seeding.

Concentration of the solution: A more super saturated solutions favors the formation of nuclei and crystals When sugar syrup boiling at 114^{0}C contains more sugar and less water than syrup boiling at 111^{0}C nuclei will form more readily from more concentrated syrup i.e syrup boiling at 114^{0}C.

Temperature: If crystallization is allowed to occur at high temperature then coarser crystals are formed. The most favorable temperature for crystal growth in sugar syrup boiling at 112^{0}C is between 70^{0}C and 90^{0}C. Suppose if temperature is lowered to 30^{0}C, the viscosity of the syrup increases and retards the crystallization

Agitation: Agitation or stirring of super saturated sugar solutions accelerates the formation of nuclei and helps in the formation of small crystals. Continuous agitation is necessary till crystallization completes, if a fine textured product is desired.

Presence of other sugars: When glucose or fructose are added to sucrose syrup, then the rate of crystallization become slow, because the glucose or fructose crystallize less readily than sucrose.

Presence of acids: When acids/acid salts are added citric acid, acid potassium tartrate, the rate of crystallization is likely to increase

Presence of invert sugar: At 30-45% level sugar will not crystallize. But when 7% added fine crystallization occurs.

Candies

Fondants-fudges-caramels and brittles

There are two types of candies-

1. Crystalline candies such as fondant and fudges; the sugar is present in the form of very small crystals and the candy feels smooth and velvety on the tongue.

2. Non-Crystalline or amorphous candies such as brittles, taffy, butter scotch and caramels. Here the sugar is not present in the form of crystals and the crystallization of sugar is prevented.

Crystalline candies

Fondant and Fudge are two types of crystalline candies. Crystallization is ensured by adjusting-

1. The consistency of sugar syrup to enable the sugar to crystallize
2. To induce crystallization by agitation
3. Addition of small amount of other ingredients which will prevent formation of large crystals ex; corn syrup, cream of tartar, butter, cream etc.

Fondant procedure

Ingredients; sugar: 400 g cream of tartar: 0.5 g corn syrup: 30 g boiling taps water: 200 ml

Procedure

1. Mix the ingredients in the cooking pan. Stir till all the sugar is dissolved.
2. Cook to the soft ball consistency 115°C in about 20 minutes.
3. Pour in to a warm shallow baking dish and cool to 40°C.
4. Beat and knead till all the lumps have broken.
5. Transfer to a jar and close with a lid

Variations in Fondant

Fondant: Is prepared from sucrose syrup boiling at 113-115°C a good fondant should be snowy white in color the crystals soft enough to be plastic and velvety but not gritty when tasted.

Variation 1 Effect of addition of egg white on fondant; the addition of beaten egg white to fondant at 3-6% level retards the growth of crystals during storage. It also imparts white color to the product.

Variation 2 Effect of hard water on the color of fondant; fondant made with a slightly alkaline hard water pH 7.2-7.8 has creamy appearance. This may be due to the traces of flavones present in the sugar which gives the color in alkaline medium. If a little cream of tartar is added to the same sugar and water a snowy white fondant is obtained. This is due to the neutralization of the alkaline salts by the acid tartrate and making the pH of water slightly acidic.

Variation 3 Effect of addition of invert sugar on fondant; fondant containing 7% reducing sugars glucose/fructose has a fine texture. If larger amounts of invert sugar are added to syrup the crystallization of fondant is prevented.

Variation 4 Addition of small amount of other ingredients; corn syrup

Variation 5 Addition of small amount of other ingredients; cream of tartar

Variation 6 Addition of small amount of other ingredients; Butter

Variation 7 Addition of small amount of other ingredients; cream

Ripening of fondant: When fondant is allowed to stand for 12-24 hours it become more moist, more plastic and kneads more easily than when it was fresh. This is known as ripening. During ripening small crystals break up thus making the mass more plastic.

Table 1. Crystallization of sugar in fondant preparations

Fondant variations	Sugar syrup boiling tempera-ture°C	Sugar syrup cooling tempera-ture°C	Formation of fondant	Formation of large crystals	Formatio of small crystals	Inferences
Basic fondant	....	....	....	....	....	....
Variation 1	....	....	....	....	....	....
Variation 2	....	....	....	....	....	....
Variation 3	....	....	....	....	....	....
Variation 4	....	....	....	....	....	....
Variation 5	....	....	....	....	....	....
Variation 6	....	....	....	....	....	....
Variation 7	....	....	....	....	....	....
Any other	....	....	....	....	....	....

Fudge procedure

Fudge: Fudge is generally prepared from brown sugar. As brown sugar contains a higher percentage of invert sugar than white sugar sucrose crystallizes less readily. Fudge ripens during storage and becomes soft and velvety after 24 hour storage.

Ingredients; sugar: 200 g Milk: 120 g butter: 14 g chocolate: 21 g

Procedure

1. Add chocolate and butter to cooking pan and heat it on a steam bath till the chocolate and butter melts. Add sugar. Mix the chocolate and butter well with sugar.
2. Then add milk and heat till the sugar dissolves completely.
3. Cook the syrup to a soft ball stage 112^{0}C.
4. Allow it to cool to about 70^{0}C and transfer to a greased molding pan.
5. Cut the candy when it is cool and wrap in butter paper or foil and store in an air tight container.

Table 2. Evaluation of Fudges

Fudge type	Appearance/setting property	Texture	Eating property	Flavor
1.	....	....	....	....
2.	....	....	....	....
3.	....	....	....	....

Non-crystalline candies

In non-crystalline candies crystallization is prevented by 1. Cooking the syrup to a high temperature so that the finished product hardens quickly before crystals can be formed. 2. Adding large amounts of materials like corn syrup, cream, butter, etc; which prevent crystallization and make the product plastic and chewy?

Procedure

Taffy

Ingredients sugar: 400 g water: 240 g cream of tartar: 0.5 corn syrup or honey: 200 g oil of lemon: 1 g

Procedure:

1. Mix the ingredients in a cooking pan. Heat and stir till the sugar is completely dissolved.

2. Cook the syrup to a hard ball stage 128^{0}C.
3. Pour quickly in to a greased plate and cut into pieces

Table 3. Evaluation of Taffy

Taffy variations	Appearance	Texture	Eating property	Flavor
1.	....	....	....	....
2.	....	....	....	....
3.	....	....	....	....

Caramels

Ingredients; sugar: 400 g corn syrup: 300 g butter: 30 g milk: 750 g

Procedure

1. Mix all the ingredients in a cooking pan and heat it till the sugar is fully dissolved.
2. Continue cooking till the syrup attains a temperature of 120^{0}C.
3. Transfer the material to a buttered pan quickly.
4. When cool, cut into pieces and wrap in butter paper. Keep caramels in air tight container.

Table 4. Evaluation of caramels

Caramel type	Appearance	Texture	Eating property	Flavor
1.	....	....	....	....
2.	....	....	....	....

Butter scotch

Ingredients; sugar: 500 g corn syrup: 125 g water: 150 g butter: 20 g salt: 1 g oil of lemon: 1 g

Procedure

1. Dissolve sugar in water and bring syrup to boil.
2. Add corn syrup and boil to 146^{0}C.
3. Add the butter in small pieces. Add oil of lemon and salt.
4. Mix and pour into a greased cooled slab.
5. While the material is still plastic cut into pieces using frame cutter.
6. Wrap in waxed foil or paper.

Table 5. Evaluation of Butter scotch

Butter scotch type	Appearance	Texture	Eating property	Flavor
1.	....	....	....	....
2.	....	....	....	....

Sugar cookery-3

Preparation of sucrose syrups for various products

Simple syrup is a solution of equal weights of sugar and water.

For example, combine 1 part of water and 1 lb of granulated sugar in a saucepan, stir, and bring to a boil to dissolve the sugar. Cool the syrup.

Dessert syrup is flavored simple syrup used to moisten and flavor some cakes. (Many chefs use 2 or 3 parts water to 1 part sugar for a less sweet syrup.)

Flavorings may be extracts, such as vanilla, or liquors, such as rum or kirsch. Add flavorings after the syrup has cooled, because flavor may be lost if they are added to hot syrup. Syrups may also be flavored by boiling them with lemon or orange rind.

Table 1. Stages of sugar cookery

Boiling temperature ^{0}C	Sugar (per cent) in syrup	Cold water test (texture of solidified syrup	Products from the syrup
110-112^{0}C	78-80	Does not form a ball	Syrups
113-115^{0}C	83	Soft ball	Fondants, fudge, burfi
118-120^{0}C	86	Firm ball	Caramels
128-130^{0}C	90	Hard ball; but plastic	Peanut candy Popcorn balls

			Puffed rice balls
138-140^{0}C	94	Very hard ball	Butter scotch
			Toffees
149-154^{0}C	98	Very hard and brittle	Peanut brittles

Procedure: Using above table prepare different recipes according to each stage of sugar cookery and evaluate it.

Table 2. Evaluation of Stages of sugar cookery and their respective sweet products

Stages of sugar cookery		Name of Product developed	Textural evaluation	Chewing capacity	Taste/taste/ flavor
Does not form a ball	110-112^{0}C	....	....	....	....
Soft ball	113-115^{0}C	....	....	....	....
Firm ball	118-120^{0s}C	....	....	....	....
Hard ball; but plastic	128-130^{0}C	....	....	....	....
Very hard ball	138-140^{0}C	....	....	....	....
Very hard and brittle	149-154^{0}C	....	....	....	....

13

Sensory Evaluation

Sensory evaluation of foods

Aim: To study different sensory testing methods using sensory evaluation score cards

Classification of Sensory testing methods

Sensory Method	No. of trained panel members	No. of samples per Test
Difference Tests		
Paired comparison test	10-May	2
Triangle test	10-May	3
Duo-Trio test	10-May	3
Rating Tests		
Ranking tests	10-May	7-Feb
Single sample test	20-Jun	1
Two sample test	20-Jun	4 pairs
Multiple sample test	20-Jun	6-Mar
Hedonic scale	12-Jun	10-May
Scoring tests		
Numerical scoring	12-May	6-Jan
Composite scoring	12-May	4-Jan

1. Sensitivity tests

Sensitivity-Threshold tests Sensitivity tests are conducted to assess the ability of individuals to detect different tastes, odors and feel the presence of specific factors in.

Procedure: You receive a series of beakers with increasing concentrations of one of the four taste qualities (sweet, salt, sour) Re-tasting of already tasted solutions is not allowed. Describe the taste and give Intensity.

Sensitivity-Threshold test Name: date:You receive a series of beakers with increasing concentrations of one of the four taste qualities (sweet, salt, sour)

Re-tasting of already tasted solutions is not allowed. Describe the taste and give Intensity

Sensitivity-Threshold test

Name :

Date :

You receive a series of beakers with increasing concentrations of one of the four taste qualities (sweet, salt, sour) Re-tasting of already tasted solutions is not allowed. Describe the taste and give Intensity

Intensity score	Description of taste
1	Weak
2	Medium
3	Strong
4	Very strong
5	Extremely strong

Use the following Intensity Scale

0 = none or the taste of pure water

? = different from water, but taste quality not identifiable

X = taste identifiable (threshold is very weak)

Sample Taste solutions (Intensity Scale)	Description of taste	Intensity score
1		
2		
3		
4		
5		

Remarks Signature

2. Paired Comparison Test

Paired Comparison Test

Name :

Product name :

Specified characteristic to be studied :

Date :

You are given one or several pairs of sample. Evaluate the two samples in the pair for ....

Is there any difference between the two samples in the pair?

Code no. of pairs	Yes	No
1		
2		
3		
4		
5		

Remarks: Signature

3. Duo-trio test:

Procedure: These test employees 3 samples, two are identical one is reference or control sample. A control sample is tested first. And two samples are judged against control, which is similar to control sample need to be identified.

Duo-Trio Test

Name :

Product name :

Date :

The first sample R is given first; Taste it carefully

Then serve two coded samples. Judge which sample is similar to R

Set. No	Reference sample	Code no. of samples	Same as 'R'
I.			
II.			
III.			
IV.			

Remarks: Signature

4. Model Triangle Test

Two of the three samples are identical. And one is odd sample

Code no.of three sample set	Sample preparations	Sample-1	Sample-2	Sample-3
ACB	Set-1	5% sugar solution	5% sugar solution	25% sugar solution
MNX	Set-2	1% salt solution	1% salt solution	10% salt solution
XYZ	Set-3	Rice Flour	Maida flour	Rice Flour
KXH	Set-4	Castor oil	Palm oil	Palm oil
DXL	Set-5	Tomato ketchup	Tomato sauce	Tomato sauce
OPT	Set-6	Groundnut chikki	Bengal gram chikki	Groundnut chikki

Triangle Test

Procedure: Prepare the three sets of samples two are identical and one odd sample just like above table and give code numbers

Name :

Product name :

Date :

Two of the three samples are identical. Determine the odd sample

Set. No & code No's		Code no of samples	Code no of odd sample	Comment on odd sample
1.	5-6-7			
2.	4-3-8			
3.	8-7-4			
4.	9-6-5			

Remarks: Signature

5. Ranking Test

Procedure: This test is used to determine how several samples differ on the basis of a single characteristic-sweetness/hardness/flavor/sourness. Panelists are presented several samples simultaneously along with code numbers. Ask the panelists to rank all the samples according to ascending/descending order

Ranking test

Name :

Product name :

Date :

Please rank the samples according to preference/ intensity of aroma/intensity of sweetness

Intensity characteristic to be studied	Code no. of samples given	Rank the samples according to intensity
Aroma/sweetness/viscosity	XX	
	YY	
	CC	
	KK	
	PP	
	M M	
	GG	
	SS	

Remarks: Signature

6. Single sample (monadic) test

Procedure: This test is useful for testing foods that have been 'after taste'. The panelists are asked to indicate the presence or absence of particular quality characteristic.

Single sample (monadic) test

Name :

Product Name :

Date :

Please taste the given and rank the samples according to detection of off-flavor in the product samples. Ex. Old pickles/ stored boondi mixture/stored sweets

Off-flavor is detected in the samples		Yes	No
If YES	Please describe it below according to intensity		
Trace			
Moderate			
Strong			

Off flavor is due to off-odor/off-taste/residual taste/other defects. Give your comments

Remarks: Signature

7. Two sample difference test

Procedure: This test measures the much of difference present between two samples in a systematic way. Prepare a pair of sample(s) with a great difference and conduct the test according to score card

Two sample difference test

Name :

Products Name :

Date :

Compare the given coded pair of samples and identify the difference between two samples in an following way

A. Whether the Difference between two samples is there or not?

There is a difference between two samples	There is no difference between two samples

B. If there is a difference between two samples, then determine the degree and direction of difference in each pair of samples on a following scale

Degree of difference	Rating Scale	Direction of difference	Directional quality
No difference	0	Superior to standard	S
Very slight difference	1	Equal to standard	E
Moderate difference	2	Inferior to standard	I
Large difference	3		

C. Taste the set of pair of samples given and interpret the results using above scale:

Pair of samples (code no)	Degree of difference	Directional difference	Remarks
100-200			
200-300			
300-400			

Remarks: Signature

8. Multiple sample difference Test

Procedure: In this test a 'standard or reference sample is served first 'R'. Taste it carefully for the quality characteristics either for flavor / aroma /odor. After that you are given several samples, which are to be compared with standard or reference sample for flavor/aroma/odor characteristics. Rate each sample according to degree of difference and directional quality given below;

Multiple sample difference Test

Name :

Products Name :

Date :

Taste the standard or reference sample first, and note down carefully for the quality characteristics either for flavor / aroma /odor of the sample served. And compare coded serving samples with standard or reference sample and rate each sample according to degree of difference and directional quality given below:

Reference sample	Code. No. of samples served	Difference from standard	Rating scale	Direction of quality
R	XXX	none	0	Equal to standard—E
R	YYY	slight	1	Inferior to standard —I
R	ZZZ	moderate	2	Superior to standard —S
R	QQQ	large	3	

Remarks on degree of difference:

Remarks on degree of direction:

Signature

9. Hedonic rating Test

Procedure: Prepare three types of recipes and display it for testing overall quality and give your opinion on the scale

Hedonic rating Test

Name :

Product name :

Date :

Taste the following coded samples given below and express 'how much you like the each product' on a following '9' point hedonic rating scale using tick mark.

'9' point hedonic rating scale	Sample code-A	Sample code-B	Sample code-C
Like extremely			
Like very much			
Like moderately			
Like slightly			
Neither like nor dislike			
Dislike slightly			
Dislike moderately			
Dislike very much			
Dislike extremely			
Reason			

Remarks: Signature

10. Numerical Scoring Test

Procedure: One or more samples are presented to each panelist in random order. The panelist evaluates each sample on a specific scale for a particular characteristic and indicate the numerical score.

Numerical Scoring Test

Name :

Product name :

Date :

Numerical Scoring Test description

Quality description	Numerical Score
Excellent	90
Good	80
Fair	70
Poor	60

Please rate these coded samples according to the above description

Code. No. of samples	Quality description	Numerical Score
FFF		
AAA		
XXX		

Remarks: Signature

11. Composite scoring test

Procedure: Sample products are displayed. Specific characteristics of a product are rated separately. This method is helpful in grading products and comparison of quality attributes by indicating which characteristic is a poor or off product. It gives more information about product than numerical score method. The panelists are trained to evaluate the dimensions of the individual quality characteristic critically.

Composite scoring test

Name :

Product name : Orange marmalade

Date :

Quality characteristics	Possible score	Sample A	Sample B	Sample C	Sample D
Color	20				
Consistency	20				
Flavor	40				
Absence of defects	20				
Total score					

Remarks: Signature

12. Descriptive Flavor Profile Test

This is both qualitative and quantitative description method for flavor analysis in products containing different tastes and odor.

Procedure: Conduct the flavor profile analysis for tomato ketchup which is prepared in the laboratory and allot score points according to the maximum scores given to the below aroma, taste, mouth feel, and texture components.

Flavor profile analysis of tomato ketchup

Aroma Components	Score	Taste Components	Score	Mouth feel Components	Score	Texture Components	Score
Garlic	1	Sour (tomato)	1	Biting sensations (Chilies, pepper etc;)	1	Tomato puree (smoothness)	3
Pepper	2	Sweet (sugar)	2				
Onion	3	Salt	1				
Cinnamon	2						
Cloves	1						

13. Dilution Test

Procedure here two samples 1. Milk and 2. Maida flour samples are given and they are diluted with other diluents factors along with maximum score for each dilution.

Prepare the similar dilution samples with any other foods. Display the samples by giving code numbers. Taste the samples and identify dilution factor and allot scores.

Dilution Test score card

Food samples	Dilution factors	Numerical score	Allotted score
Pure milk	Without dilution	5	
1. Pure milk diluted with milk powder	Addition of milk powder	4	
2. Pure milk diluted with 25% water	Addition of least amount water	3	
3. Pure milk diluted with 50% water	Addition of maximum water	2	
Pure Maida	Without dilution	5	
1. Pure Maida diluted with 25% ragi flour	Addition of least amount of ragi flour	4	
2. Pure Maida diluted with 50% ragi flour	Addition of moderate amount of ragi flour	3	
3. Pure Maida diluted with 75% ragi flour	Addition of maximum amount of ragi flour	2	

14. Dilution Flavor Profile Test

Procedure: Prepare any recipe with complete ingredients and other recipes prepare with one missing flavor ingredient each time display for tasting and identification

Recipe	Components	Flavor Dilution levels	Numerical score
Meat curry	Meat +condiments and spices salt and chill powder + cashew paste + water to cook	No dilution / good flavor	100
	Meat +condiments and spices salt and chill powder + water to cook	Dilution due to removal of cashew paste	80
	Meat +condiments and spices salt and chill powder + water to cook + addition of tomatoes		
	Meat +condiments and spices salt and chill powder + water to cook + addition of tomatoes + addition of large quantity of water	Dilution due to addition of tomatoes	40
	Meat + salt and chill powder + addition of tomatoes + addition of large quantity of water	Dilution due to addition of extra amount of water	60

14

Bakery Products

Aim: To learn the preparation of various types of bakery products and evaluate their characteristics

Bakery products 1- Bread preparation methods

Formulation & method of bread making Baking is an exact science that requires precise measuring and accuracy. Accuracy is crucial in baking because most baked products are made from the same basic ingredients: flour, liquid, fat, sugar, salt etc. the difference between baked products are often the proportion of each ingredients in the formula. If the proportions are off, you will end up with a different product or an unacceptable product. The bakers' percentage allows you to change the yield of a formula without changing the quality of final product. A bakers' percentage means that each ingredients is a certain percentage of the weight of the total flour in the formula. Convert the following recipe into bakers' percentage and formula percent.

Table 1. Method of Basic bread

Ingredients	Weight g	Baker's formula percentage
Flour	500	
Water	300	
Salt	10	
Yeast	5	

$$\textbf{Baker's percentage} = \frac{\text{Weight of ingredient}}{\text{Total weight of the flour}} \times 100$$

Bread formulas

Table 2. Proportions of ingredients used in bread formulations

Ingredients	White bread—I	White bread—II
Flour (hard wheat)	100	100
Water	60.6	60.6
Yeast	1.8	1.8
Salt	2	2
Milk dry (skim)		5

Table 3. Formulations for straight dough and sponge dough breads

Ingredients	Straight dough bread	Sponge-dough bread	
		Sponge	Dough
Flour	100	65	35
Water	65	45	20
Yeast	3	2.5	—
Yeast food	0.2-0.5	0.2-0.5	—
Salt	2.25	—	2.25
Sugar	8	—	8
Fat	3	—	3
Non-fat milk solids	3	—	3

Steps in Preparation of bread

1. Mixing of ingredients and preparation and kneading of dough till it develops smooth surface and rest for 15-20 minutes so that gluten develops well
2. Fermentation of the dough
3. Baking of bread dough; bread is baked at 226.7°c for about 30 min. as the dough enters the hot oven maintained at 226.7°c, a visible film forms on the dough surface. Next, the expansion of the dough volume to the extent of 30% occurs.
4. Cooling and packing; cooling of bread is essential in order to avoid moisture condensation before packing. Packing of bread should be done in moisture proof films or wrapping in waxed paper to prevent loss of moisture from crust and crumb.

Table 4. Evaluation of bread preparation

Bread types	Aeration of dough before fermentation (volume of dough) ml	Aeration of dough after fermentation (volume of dough) ml	Taste of bread after baking	Crust structure characteristics Thick, dry/ or soft, moist	Crumb structure characteristics Soft & spongy / or sticky crumb
1.					
2.					
3.					
4.					
5.					

Bakery products 2- Biscuits preparation methods

Table 5. Proportions of ingredients used in biscuit formulations

Ingredients	Sweet biscuits	Salt biscuits	Protein enriched
Soft wheat flour	100	100	100
Shortening(fat)	10	10	10
Cane sugar	20	5	20
Salt	……	0.6	……
Skim milk or powder/ peanut flour/soy bean flour	…….	…….	……..
Sodium bicarbonate			
Ammonium carbonate			
Flavor	0.4	0.4	0.4
Calcium phosphate	0.2	0.2	0.2
Vitamin premix	0.1	1.2	1.2
	……	…….	0.5
	…..	…….	1.2

Steps in biscuit preparation

Mixing and kneading: Dry ingredients like flour and sugar are transferred to a mixing vessel, whereas shortenings and flavorings are added and mixed in a mechanical mixer. Then water and baking powder are added to mixer and continue mixing till a desired consistency of dough is obtained.

Sheeting and shaping: The dough is rolled into sheets of desired thickness by passing it through pairs of rolls. The rolled dough sheet is cut by a mechanically worked stamping divider fitted with dies.

Baking and cooling: The cut biscuits move forward on a continuous belt from which they are automatically transferred to a continuous steel plate or wire mesh bands travelling through the ovens. The length of the ovens will depend on the production capacity. The biscuits are baked at 232.2^{0}c for 15 min and cool the biscuits

Packaging: The biscuits should be packed in tins or moisture proof packaging. The moisture proof and grease proof glassine, cellophane and metal foil laminated moisture proof papers extend the shelf life of biscuits for long periods for 6 months without the development of off-flavors.

Table 6. Evaluation of biscuits preparation

Biscuit types	Baking performance-able to spread the dough	Diameter of biscuit (cm)	Brittleness of biscuit-crispy/ crunchy/	Dry or moistness biscuit	Taste/ flavor
1					
2					
3					
4					
5					

Bakery products 3- Cakes preparation methods

Proportions of ingredients used in various types of Cake formulations

Batter Type Cakes Batter type cakes can be divided into (i) High ratio cake and low ratio cake. High ratio cake is that which contain more sugar than flour while low ratio contains less sugar than flour. Basic differences between them are the stiffness or fluidity of the batter. High ratio cake in view of high sugar content requires more water to dissolve the sugar. Introduction of emulsified shortening enables to use high amount of water. Because of flow nature of batter, the capacity to retain gas is very low. Hence, chlorination helps flour to retain gas due to faster "set" of cake in the oven.

Formula balance for low ratio and high ratio cakes are given below

Low ratio

1. Sugar not more than flour
2. Total liquid = liquid in egg and milk
3. Total liquid = sugar
4. Shortening not to exceed egg

High ratio

1. Sugar greater than flour
2. Total liquid = liquid in egg and milk

3. Total liquid is more than sugar
4. Egg equal or greater than shortening

Table 7. Formula Batter type of cakes

Ingredients	High ratio	Low ratio
Flour	100	100
Sugar	100-180	80-98
Shortening	54	50
Egg	60	55
Salt	4-Feb	2-Jan
Baking powder	4-Mar	2-Jan
Milk	100	60
Flavor	As desired	As desired

Foam Type Cakes: These are class of cakes which are leavened by air and whipped into egg portion. There are 2 types (i) Angel food and (ii) Sponge cake. Angel food cakes make use of egg white to entrap air while sponge cake makes use of mixture of whole egg and egg yolk.

Table 8. formula of Angel food cake

Ingredients	Amount
Flour	100
Sugar	260-300
Egg white	260-300
Acid salt	4 to 4.5
Salt	4 to 4.5
Flavor	As desired

Acid salt is used to neutralize the alkalinity of egg white and also to strengthen the egg. Iffruitjuices and acid fruits are used, acid salt can be reduced. 60-70% of sugar could be used in the 1si stage of mixing. Remaining sugar when added with flour enables distribution of flour more evenly. In these types of cakes sugar is equal to egg white. Fruits and nuts could be added equal to weight of flour. Lemon rind, oranges if used, moisture content and acidity should be considered.

Sponge cake: Sponge cakes are of 2 types one is a basic sponge which contains flour, egg mixture, sugar and salt. Short sponge in addition to the above ingredients, contains additional sugar, water, milk solids and baking powder.

Table 9. formula of basic Sponge cake

Ingredients	Amount
Flour	100
Sugar	166
Egg mixture	166
Salt	3

Pound cake: Pound cake represents oldest example of aerated fat containing cake.

Table 10. Formula of Pound cake

Ingredients	Amount
Flour	450 gm
Butter	450 gm
Egg	450 gm
Sugar	450 gm
Lemon extract'	To taste

Table 11. formula of fruit cake

Ingredients	Level	Fruit	Level
Flour	100	Glaced cherry	75 g
Sugar	85	Orange peel	25 g
Fat	65	Tuityfiuity	75 g
Eggs	100	Cashew nuts	75 g
Baking powder	1 g		
Vanilla	1 ml		
Caramel	2 ml		

Cakes preparation method

Conventional Mixing method: Sugar is mixed well with fat, and then whole egg is then added to the sugar-fat mixture and mixed well with uniform consistency and fluffy mass. Sift/sieve the flour along with baking powder and salt. Milk and flavorings are combined. Small quantities of flour and milk are combined and added alternatively to the fat-sugar-egg blend. The blend should be mixed after each addition. Mix the blend well till a uniform batter is obtained.

Quick mixing method: All the dry ingredients flour, sugar, baking powder and fat are mixed together. The flavoring is mixed with milk and added to the flour blend. The blend is mixed well in an electric mixer for 2 min. Eggs are then added to the blend and mixed again for 2 min in an electric blender.

Sugar batter method

The fat and sugar are creamed together until light and fluffy. The warmed egg is added in intervals (small additions) into the mixture, ensuing that with each addition the fat mixture is well creamed and not separated. The conditioning of the egg is very important as curdling of the batter can occur at this stage, mostly due to too cold egg. Curdling is the breakdown of the emulsion, which is being formed, as the fat separates out from the liquid. The egg should be warm, but these are the consequences if the temperature is incorrect:

- Egg too cold – the fat hardens, air escapes and the mix curdles
- Egg too warm – the fat turns to oil, the air escapes and the mix curdles
- Egg added too fast – the mix becomes saturated, the air escapes and the mix curdles.

It may also be possible to add the eggs in a steady stream; care must be taken not to curdle the mixture. The batter should have a soft and velvety texture, after all egg is added. The flour is sifted and gently mixed through the batter, until it is clear and smooth. Do not over mix, as this would cause toughness.

Flour batter method

The fat is mixed with one third of the sifted flour until it is well creamed (+/-8 minutes), ensuring that the entire batter is aerated by scraping the bowl down. The egg and sugar is whisk to a foam (sponge), using a separate bowl. Egg and fat need to be of the same temperature and consistency before they are combined; add some of the egg mixture into the fat to adjust consistency. Carefully fold into the fat mixture the following: remainder of eggs, sifted flour and baking powder and lastly the liquid. Each ingredient needs to be cleared in the batter, before adding the next ingredient. In order to avoid any lumps it is vital to follow the sequence.

The Blending method

The Blending Method does not require aeration or creaming of the fat with the sugar or the flour. The aeration of the batter takes place towards the end of the mixing cycle, rather than being the first step, as in the sugar or flour batter methods.

Sponge Production methods

All weighing and mixing equipment must be free from grease, wash with hot soapy water and rinse prior to use.

Traditional method: (Orthodox Sponge)

A basic egg sponge with ingredient ratio of 2 parts egg; 1 part sugar; 1 part flour. The light texture is obtained by whisking the eggs and sugar together on a high speed, with the flour carefully folded in last. Nowadays some formulas contain small amounts of baking powder and also can contain some butter. Egg and sugar is warmed to 38°C, which increases the foam stability, due to the egg protein. Whisk on top speed until a full foam (sabayon) is achieved. If the mixture is then whisk for a short time on medium speed it will produce better and more stable foam, which in turn produces a better sponge. The sifted flour is then carefully folded into the batter, ensuring not to lose the trapped air. Immediately the batter is filled into baking dishes and baked for best results. Delays in baking of sponges often result in air loss and poor volume.

Enriched method: (Genoese Sponge)

The Genoese sponge is the same as an orthodox sponge, but it contains fat (up to 80% of the sugar weight), this addition increases shelf-life and handling properties of the sponges. Melted butter (+/- 30°C) is folded into the traditional sponge after the sifted flour is incorporated. Ensure cooler temperatures for butter, if too hot the egg will curdle.

Emulsified/stabilised sponges

The type more widely produced is the stabilised or emulsified sponge which differs greatly in mixing technique and handling properties. For this formula, an all-in method is used, and as the name suggests, there is the addition of a stabiliser or emulsifier. Unlike the orthodox sponge, this sponge can be made and stored, which is due to the stabilised emulsion which retains its condition. The recipes usually contain water and proportions of baking powder. Emulsifiers and stabilisers are available in powder or paste forms, the majority of them are based on lecithin and lacto albumen, these emulsifiers enable normally incompatible substances such as water and the fat from the egg yolk to combine and form an emulsion. The action of the emulsifier is assisted by the beating process which reduces the egg particles to the same size as those of water. The result is more evenly distribution of fat and water.

Baking: Once the batter is prepared according to above methods the batter is transferred to cake pans. The pans are kept in a baking oven set at 190^{0}c and baked for 30 minutes.

Evaluation of cakes

Table 12. Evaluation of cake preparation

Cake type	Creaming performance- fluffy/thick/ thin mass	Consistency of batter before baking thick/thin	Volume of cake after baking	Well cooked/ under cooked	Crust structure charact- eristics	Crumb structure charact- eristics
1						
2						
3						
4						
5						
6						

Check the quality characteristics of the prepared cakes according to the factors given in table 6

Table 13. Causes of variation in cake quality

Variations in cake	Causes
Streaks on top of cake	1. Under baking 2. Cake moved during baking 3. Too hot oven
Streaks on bottom of cake	1. Too much liquid 2. Insufficient baking powder 3. Insufficient sugar 4. Too soft flour 5. Insufficient egg
Collapse in centre of cake and white spots on the crust	1. Excess sugar
Collapse in centre of cake-dark crust	1. Excess baking powder
Small volume-collapse at sides, shrinking from sides	1. Excess liquid 2. Insufficient egg 3. Too soft flour
Small volume-cauliflower top	1. Too hot oven 2. Insufficient steam in oven

	3. Too strong flour 4. Too much egg 5. Insufficient sugar
Long holes in texture	1. Insufficient aeration 2. Insufficient creaming 3. Poor creaming fat 4. Too much mixing
Discolored crumb	1. Badly balanced baking powder-excess alkali
Too tender crumb	1. Too much fat in relation to egg
Crumbly crumb-coarse open texture	1. Weak flour 2. Insufficient egg in relation to fat 3. Too much sugar 4. Too much baking powder 5. Slow baking
Shrunken appearance	1. Insufficient flour and weak flour 2. Insufficient egg 3. Over creaming 4. Insufficient mixing 5. Slow baking 6. Fruit too heavy 7. Very wet fruits 8. Too much sugar 9. Too much baking powder

Bakery products 4- Pastries preparation methods

Pastry is a dough of flour, water and shortening (solid fats, including butter) that may be savory or sweetened. Sweetened pastries are often described as bakers' confectionery. Small tarts and other sweet baked products are called pastries. Common pastry dishes include pies, tarts, quiches, croissants, and pasties

Ingredients required: Flour, sugar, milk, butter, shortening, baking powder, eggs and other optional ingredients

Types of pastries

Short crust pastry

Short crust pastry is the simplest and most common pastry. It is made with flour, fat, butter, salt, and water to bind the dough. This is used mainly in tarts. The process of making pastry includes mixing of the fat and flour, adding water, and rolling out the paste. The fat is mixed with the flour first, generally by rubbing with fingers or a pastry blender, which inhibits gluten formation by coating the gluten strands in fat and results in a short (as in crumbly; hence the term short crust), tender pastry, in which sugar and egg yolks have been added (rather than water) to bind the pastry.

Flaky pastry

Flaky pastry is a simple pastry that expands when cooked due to the number of layers. It bakes into a crisp, buttery pastry. The "puff" is obtained by the shard-like layers of fat, most often butter or shortening, creating layers which expand in the heat of the oven when baked.

Puff pastry

Puff pastry has many layers that cause it to expand or "puff" when baked. Puff pastry is made using a laminated dough consisting of flour, butter, salt, and water. The pastry rises up due to the water and fats expanding as they turn into steam upon heating. Puff pastries come out of the oven light, flaky, and tender.

Pastry formulations

For 8-Inch Shell:

- Ingredients required
- 1 c. plus 2 tbsp. flour
- 1/4 tsp. salt
- 1/3 c. shortening or butter-flavored shortening
- 2 to 3 tbsp. cold water

For 9-Inch Shell:

- Ingredients required
- 1 1/2 c. flour
- 1/4 tsp. salt
- 1/2 c. shortening
- 3 to 4 tbsp. cold water

For 9-Inch, 2 Pie Crusts:

- Ingredients required
- 2 1/2 c. flour
- 1/2 tsp. salt
- 3/4 c. shortening
- 6 to 7 tbsp. cold water

Method of preparation:

1. Mix flour and salt.
2. Cut shortening with a pastry blender or fork. Mixture should be size of small peas.
3. Its texture will not be uniform, but will contain crumbs and small bits and pieces.
4. Sprinkle water over flour mixture a tablespoon at a time.
5. Mix lightly with fork, using only enough water so that the pastry will hold together when pressed gently into a ball. Roll dough out and place in pie pan.

Table 14. Evaluation of pastries

Type of pastry/ variation	Texture out come	Hard pastry/ tender pastry	Flakiness of pastry	Eating property
1				
2				
3				
4				
5				
6				

15

Food Preservation Technologies-Drying of Foods

Aim: To learn the drying and dehydration practices of fruit and vegetables and to evaluate their characteristics

Principle: Drying is one of the oldest methods of preserving food. It is usually accomplished by the removal of water. Dried foods are preserved for long storage time because the available moisture is very low, so that the micro organisms cannot grow and enzyme activity is controlled. It is done by many methods. 1. Sun drying 2. Using mechanical dryers

Food preservation by Sun drying method

Sun drying is done by drying food under the sun rays, ex; grains and nuts. Sun drying is a slow process, risk of contamination and spoilage and is limited to climates and environment hot sun and dry atmosphere. The types of fruits suitable for sun drying are grapes figs, dates, apricots, raw mangoes. And vegetables such as beans, peas, cabbage, cauliflower, lady finger, onions, chilies, turmeric, green leafy vegetables.

Processing steps involved in Sun drying of Vegetables

Method of preparation:

1. The fruit/vegetables are washed, peeled and cut into thin slices
2. The prepared F/V slices are spread on the flat bottom trays
3. Drying under the hot sun till desired dryness
4. Dried F/V slices are obtained
5. Stored under air tight container

Sun drying of papads

Method of preparation:

1. Take Black gram flour +black pepper, salt + vegetable oil

2. Mix all the ingredients and kneaded to form the dough
3. Take a small dough and flatten it and cut into circles shape
4. Spread on the plastic or cotton sheet
5. Dry under the hot sun
6. Dehydrated papads are obtained
7. Stored under airtight containers

Sun drying of Vadiyalu

Method of preparation:

1. Prepare the rice flour batter
2. Take water in a pan and bring to boil
3. During boiling of water add cumin seeds + salt + prepared rice batter slowly
4. Keep stirring till it get cooked to desired consistency
5. Spread either plastic sheets/ or cotton cloth on the area where sun rays fall adequately
6. Spread the batter with the help of the spoon on the spreader sheet as a thin layers
7. Dry for 3-4 days under the hot sun till all the material get completely dried uniformly
8. Collect the dried vadiyalu and store in a airtight containers

Table 1. Evaluation of Sun dried foods

Type of food sun dried	Vegetables /texture	Fruits /texture	Vadialu /texture	Papads /texture	Dehydration ratio= Initial weight- final weight x 100	Total weight of dehydrated sample per kg of fresh sample
1						
2						
3						
4						
5						

Food preservation by Dehydration Technology

Drying is one of the oldest methods of food preservation. Drying preserves foods by removing enough moisture from food to prevent decay and spoilage. Water content of properly dried food varies from 5 to 25 percent depending on the food.

Successful drying depends on

- Enough heat to draw out moisture, without cooking the food;
- Dry air to absorb the released moisture; and
- Adequate air circulation to carry off the moisture.

When drying foods, the key is to remove moisture as quickly as possible at a temperature that does not seriously affect the flavor, texture and color of the food. If the temperature is too low in the beginning, microorganisms may survive and even grow before the food is adequately dried. If the temperature is too high and the humidity too low, the food may be harden on the surface. This makes it more difficult for moisture to escape and the food does not dry properly. Although drying is a relatively simple method of food preservation, the procedure is not exact. A "trial and error" approach often is needed to decide which techniques work best.

Dehydration of Vegetables

Processing steps involved in Dehydration of Vegetables

1. Selection of drying trays

Drying trays can be simple or complex purchased or built. Good air circulation without reaction between food and trays is most important.

For small amounts of food and trial runs, cheesecloth or synthetic curtain netting stretched over oven racks, cake racks, broiler racks or cookie sheets work well. Attach with clothes pins.

For large quantities of food use shallow wooden or plastic trays with slatted, perforated or woven bottoms.

If preparing your own trays do not use galvanized screening for tray bottoms. It has been treated with zinc and cadmium, which can cause a harmful reaction when in contact with acid foods.

Other metals such as aluminum also are not advisable because they may discolor and corrode with use. If used, line with cheesecloth or synthetic curtain netting to keep food from touching the metal. A liner also helps keep foods from sticking to trays and prevents pieces of food from falling through.

Wash trays in hot, sudsy water with a stiff brush. Rinse in clear water and air dry thoroughly before and after each use. A light coat of fresh vegetable oil or nonstick substance helps protect wood slats and makes cleaning easier. If trays are used in an oven, they should be 1 1/2 inches smaller in length and width than the oven dimensions to allow for good air circulation.

2. Selecting Vegetables for dehydration

Select vegetables at peak flavor and eating quality. This usually is just as they reach maturity. Sweet corn and green peas, however, should be slightly immature so they retain their sweet flavor before their sugars change to starch.

Picking activates enzymes that cause color, flavor, texture, sugar content and nutrient changes in vegetables. To control such changes, prepare the produce immediately after gathering and begin processing at once. Thoroughly wash or clean produce to remove any dirt or spray. Drain thoroughly. Shake leafy vegetables well. Sort and discard any food with decay, bruises or mold. Such defects may affect all pieces being dried.

3. Pre-treating Vegetables

To enhance Quality and Safety Pre-treating vegetables by blanching in boiling water or citric acid solution is recommended to enhance the quality and safety of the dried vegetables. Blanching helps slow or stop the enzyme activity that can cause undesirable changes in flavor and texture during storage. Blanching also relaxes tissues so pieces dry faster, helps protect the products vitamins and color and reduces the time needed to refresh vegetables before cooking. In addition, research studies have shown that pre-treating vegetables by blanching in water or citric acid solution enhances the destruction of potentially harmful bacteria during drying, including Escherichia coli O157:H7, Salmonella species and Listeria monocytogenes.

4. Blanching the vegetables

Water blanching is recommended over steam blanching or blanching in a microwave because water blanching achieves a more even heat penetration than the other two methods. Plain water or water with added citric acid may be used. Citric acid acts as an anti-darkening and anti-microbial agent. Prepare the citric acid water by stirring 1/4 teaspoon (1 gram) of citric acid into one quart (approximately one liter) of water. Work with small amounts so plain or citric acid water doesn't stop boiling.

Watch closely and precook as follows

- Fill large kettle half full with plain or citric acid water and bring to a boil.
- Put no more than one quart of the vegetable pieces in a cheesecloth or other mesh bag. A 36-inch cloth square gathered at the corners works well. Secure ends.
- Drop vegetable bag in boiling water, making sure water covers the vegetables. Shake bag so hot water reaches all pieces.
- Start timing as soon as vegetables are in boiling water. Adjust heat to ensure continuous boiling.
- Drop bag in very cold water to cool (same time as blanched).
- Drain on paper towel or cloth.

5. Drying methods for dehydration of vegetables

Arrange pretreated vegetables on drying trays in single or thin layers, 1/2 inch deep or less. Dry in dehydrator or oven as described below.

Thermostatically controlled electric dehydrators are recommended for home food drying. They are relatively inexpensive, convenient for drying large or 0small batches of food, and easy to use. The best dehydrators have thermostatically controlled heat settings and fans that blow warm air over the foods. Some models have a heat source at the bottom and removable, perforated trays (for air circulation) stacked above the heat source. Dehydrators should be used indoors in a dry, well-ventilated room. Food on lower trays near the heat source will often dry more rapidly than food on higher trays and, therefore, trays should be rotated throughout drying.

Oven Drying If you do not have access to a food dehydrator, either a gas or electric oven may be used to dry vegetables. Both require careful watching to prevent scorching. Proper temperature and ventilation are most important in oven drying. To oven dry, preheat oven at lowest setting (140 to 150 degrees F), then adjust the thermostat and prop the oven door open to achieve a consistent oven temperature of 140 degrees F, and to allow moist air to escape.

Conventional ovens may not maintain consistent temperatures at low settings. To ensure maintenance of 140 to 150 degrees F, monitor oven temperature using a calibrated oven thermometer. Place the oven thermometer directly on the oven rack or tray and check it every two hours throughout drying. Place trays of prepared food in oven. Stack trays so there is at least 3 inches of clearance at the top and bottom of the oven and 2 1/2 inches between trays. Shift trays, top to bottom and front to back, every half hour and stir food often if it is 1/2-inch deep or more. Single layers need no stirring. Food scorches easily toward the end of drying time; therefore, turn the heat off when drying is almost complete and open the door wide for an additional hour or so.

6. Post-Drying Treatment

When drying is complete, some pieces will be moister than others due to size and location during drying. Conditioning distributes residual moisture evenly in dried food. In doing so, it reduces the chance of spoilage. Because vegetables dry to a nearly waterless state, conditioning them is not always necessary.

7. Packaging and Storing

Pack cooled dried foods in small amounts in dry, scalded glass jars (preferable dark) or in moisture- and vapor-proof freezer containers, boxes or bags. Metal cans may be used if food is first placed in a freezer bag. To protect from insects and re-absorption of moisture, seal lids onto containers. Wrap the edge where the lid meets the container with a plasticized, pressure sensitive tape or clean, 1-inch cloth strip dipped in melted paraffin. Bags may be heat-sealed or closed with twist ties, string or rubber bands. Label containers with the name of the product, date, and method of pretreatment and drying and store in a cool, dry, dark place. Properly stored, dried vegetables keep well for six to 12 months. Discard all foods that develop off smells or flavors or show signs of mold.

Table 2. Evaluation of dehydrated vegetables

Name of vegetable	Weight of Fresh vegetable gm or /kg	Blanching time (minutes)	Drying time (hours)	Dryness test/brittle/ very dry	Weight of dried vegetable gm/kg	Yield of the product g/ or kg
1.						
2.						
3.						
4.						
5.						
6.						

16

Food Preservation Technologies Processing of Fruit Products

Aim to learn the different preservation techniques of fruits and vegetables and evaluate their characteristics

Processing of Jam

Steps involved in processing of Jam

1. **Selection fresh fruits**/frozen stored fruits/chilled fruits/fruit pulp/sulphited fruit. Wash them to remove the dairt, dust, stems, leaves, stalks, undesirable portions.

2. Preliminary treatment

- For strawberries—crushing the fruit between rollers
- For ras berries —steamed, crushed, passed through sieves to remove cores
- For plums and cherries —heating with water till they become soft and are then passed through sieves to remove stones
- For Goose berries are whirled in a machine line with carborundum to rub off the tops and tails, passed through sieves to remove stalks.
- For pears—peeling, coring, cut into small pieces
- For Apricots—cutting, remove the stones
- For Mangoes— peeling, coring, remove the stones, cut into small pieces
- For Apples— peeling, coring, cut into small pieces
- For Grapes— heating with water till they become soft and are then passed through sieves to remove seeds
- For pineapple— peeling, coring, remove the eyes, pitting

3. Preparation of fruit pulp

After pretreatment is completed, the obtained fruit pulp is taken and subjected to pass through sieves to remove any adhering fibers, crushed seeds of minute nature, and fine peel particles. Now the fruit pulp is ready and should be placed in boiling pans.

4. Addition of Sugar

Add 100 ml of water to 1 kg of fruit pieces, cook for 10 min and crush the fruit pieces with a ladle. Add 1kg sugar and 2 g citric acid. Generally a fruit pulp to sugar ratio of 1:1 is followed. If the fruits are acidic nature, the fruit pulp to sugar ratio of 1:2 is followed. Mix thoroughly all the ingredients.

5. Addition of Acid

Generally, citric, malic, tartaric acids are used to supplement the acidity to the low acid fruits as well as addition of acid is necessary because, appropriate combination of pectin-sugar-acid ratio is essential in order to set the jam into desired/sheeting consistency. The PH of the fruit juice should be 3.1-3.5 before the sugar is added.

6. Addition of colors and flavors

Only permitted edible food colors and flavors are used. And these should be added at the end of boiling process

7. Cooking /boiling process

Cook the mixture slowly with occasional stirring. The fruit pieces should be crushed with ladle during cooking. Continue cooking till the temperature of the mass reaches 105.5^{0}c.

8. Judging of end point

a) **Sheet test** Take the product in a spoon and allow it to flow. If it falls as a sheet, then the jam is ready. If the product flows, then cooking should be continued for some more time till the sheeting test is positive.

b) **Drop test** Take the product in a spoon and allow it to cool and add a drop of it in a glass filled with water. If the drop settles down at the bottom of water without any disintegration, denotes the end point.

c) **Refract meter method** This method is commonly followed by large scale food industries. The cooking is stopped when refract meter reading shows 69^{0}Brix.

d) **Boiling point method** when jam containing 69% TSS (total soluble solids) boils at 106^{0}c at sea level.

9. Packaging

Fill the hot jam into clean dry and sterile jars or cans and allows to cooled and fix the sterilized lid to the jar/ seal the cans, followed by labeling and packaging, and store in a cool place.

Table 1: Evaluation of jam

Fresh weight of fruit pulp	Jam setting/ sheeting	Refract meter reading (^{0}Brix) test	TSS %	Color/ appearance of jam	Spread ability on bread/ roti slices	Taste
1.						
2.						
3.						
4.						

Processing of jellies

A jelly should be transparent and well set and should have the original fruit flavor. It should keep to shape when removed from the mould. The final product should have 65 percent solids and 45 g of fruit extract in 100 g of jelly.

Fruits for jelly should be sufficiently ripe mature but firm but not over ripe, and should have good flavor and contain sufficient acid and pectin to yield a good jelly, without the addition of pectin and acid.

Slightly under ripe fruits yield more pectin than over ripe fruits. Because in over ripe fruits, the pectin decomposes into pectic acid, which does not form a jelly with acid and sugar, hence over ripe fruits should not be used for jam and jelly making. Hence jelly is prepared by combining slightly under-ripe fruit extracted pectin with-acid-and -sugar in appropriate proportions.

The steps involved in preparation of jelly from guava

1. Selection and preparation of fruit: Good quality jelly can be prepared from guava, apple, grapes, papaya, jaman, and jackfruit. Select sound slightly unripe guava fruits. Wash them well in cold water and cut into small pieces with a stainless steel knife.
2. Extraction of pectin: Take boiling pan and fill the 1 kg fruit, cover with 1½ liters of boiling water. Add 2 g citric acid. Boil the mass crushing the fruit well with the ladle for 30-35 min. strain the mass through a coarse cloth. Keep the extract in a tall container to allow the solids to settle. Decant carefully the clear extract.

3. Test for pectin: To a spoon of extract, add two teaspoons of methylated spirit. Formation of one big clot indicates 'high pectin content'; formation of thin gelatinous precipitate indicates 'low pectin content'.
4. Addition of pectin: if pectin content is low, then 2.5 g of pure pectin should be added for 500 g sugar
5. Addition of sugar and Acid: to 500 g of extract, 500 g of sugar and 1 g citric acid are added
6. Cooking: the mixture is cooked till the boiling point is increased to 105.5^{0}c and it gives sheeting test. The importance of boiling is, during boiling it dissolves the sugar acid pectin and bring union of pectin-acid-sugar to form jelly and jelly strength.
7. Sheeting test: It is carried out by taking the cooked material in a spoon and it is allowed to flow. If it falls in the form of a sheet then the end point is reached. If it falls in thick syrup, it requires further cooking till the setting point /boiling point reaches 105.5^{0}c.
8. Pour the hot jelly into clean dry sterilized glass jars and fix the lid. Aloe the product to cool and seal the lid airtight.

Table 2: Evaluation of jellies

Fresh weight of fruit pulp	Jelly setting /sheeting test	Refract meter reading (^{0}Brix)	TSS %	Jelly strength	Jelly quality	Color/ appearance of jam	Spread ability on bread/roti	Taste
1.								
2.								
3.								
4.								

Processing Mango nectar

1. Washing, peeling, and slicing.
2. The slices are passed through pulping machine to extract the pulp
3. Take Mango pulp-3 kg, sugar-3 kg, citric acid- 30 g and water- 5 kg
4. The mixture is heated to 100^0 C and passed through a homogenizer

5. The homogenized and the heated juice is filled in the heated cans at 100°C and double seamed, sterilized at 100°C for 30 min. and cooled in running water

Processing of fruit syrups

Extracted clear fruit juice -------- 1 kg

Sugar -------- 1 kg

Citric acid -------- 15-20 g

The mixture is heated at 100°C for 30 min. to dissolve the acid and sugar. As a result Fruit syrup is produced. The syrup is poured into heated bottles at 85°C to 87.8°C, fitted with crown cork and pasteurized at 79.4°C for 30 min.

Processing of Pure fruit juices

Grape juice and pineapple juice

1. The juice is extracted by crushing the grapes in a basket press.
2. Then the juice is filtered through cloth
3. The filtered juice is heated to 85°C for 15 min.
4. The heated juice is filled in the heated bottles and sealed
5. The sealed bottles are subjected to pasteurized at 82.2°C for 30 min. by the 'overflow' method

Apple juice

1. Wash the apples in dilute HCL (5 lit acid in 100 lit of water) to remove the arsenic and lead spray residues and followed by wash with water
2. The washed apples are cut into small pieces with apple grater
3. The juice pressed out with hydraulic press and collected in stainless steel vessels
4. The juice is filtered through coarse cloth
5. The filtered juice is heated to 85°C for 15 min
6. The heated juice is filled in the heated bottles and sealed
7. The sealed bottles are subjected to pasteurized at 82.2°C for 30 min. by the 'overflow' method without allowing in air spaces

Processing of fruit Squashes

Fruit squashes should contain not less than 25 per cent fruit juice and 40 % TSS (total soluble solids) and it should contain added citric acid, sucrose and preservatives

Method of preparation

1. The juice is extracted from fruits and keep aside
2. Sugar and citric acid are dissolved in hot water to thick syrup and filtered through a cloth.
3. The syrup is mixed with fruit juices and should boil till it gets TSS of 40% and 25% of juice and 0.8-1.0 per cent Acidity
4. Add edible colors and preservatives 350ppm of SO_2 (sulphur dioxide) or 600ppm of benzoic acid and thoroughly mix with the syrup containing juice and cooled
5. The squashes are filled in clean sterilized bottles and sealed with crown cork
6. The bottles are stored under following conditions;
   a. 38^{0}C and 92% RH or
   b. 27^{0}C and 65% RH or
   c. Refrigerated temperature 4^{0}C

Table 3: FPO specifications for fruit beverages

Type of beverage	Minimum % of TSS (total soluble solids) in final product	Minimum % of fruit juice in final product
1. Fruit squash	40	25
2. Fruit syrup	65	25
3. Cordial	30	25
4. Sweetened juice	10	85
5. Fruit juice concentrates	32	100

Table 4: Permissible limits of preservatives in fruit beverages

Fruit beverage	Preservative	Max. level permitted(ppm)
1. Fruit juice concentrates	Sulfur dioxide	1500
2. Squashes fruit syrups, cordials, fruit juices	Sulfur dioxide or	350
	Benzoic acid	600

Table 5: Evaluation of fruit products

Weight of the fruit pulp (gms)	Juice quantity (ml) in final product	TSS% in final product	Percent dilution required	Taste/color
Fruit squash				
Fruit syrup				
Cordial/nectars				
Sweetened juice				
Fruit juices				
Fruit concentrates				

17

Food Preservation Technologies Processing of Pickles

Aim: To learn to prepare different types pickles and evaluate the factors in it

Pickling is the process of preserving or extending the shelf life of food by either anaerobic fermentation in brine or immersion in vinegar. The pickling procedure typically affects the food's texture and flavor. The resulting food is called a pickle, or, to prevent ambiguity, prefaced with pickled. Foods that are pickled include vegetables, fruits, meats, fish, dairy and eggs.

A distinguishing characteristic is a pH of 4.5 or lower, which is sufficient to kill most bacteria. Pickling can preserve perishable foods for months. Antimicrobial herbs and spices, such as mustard seed, garlic, cinnamon or cloves, are often added. If the food contains sufficient moisture, pickling brine may be produced simply by adding dry salt. For example, sauerkraut and Korean kimchi are produced by salting the vegetables to draw out excess water. Natural fermentation at room temperature, by lactic acid bacteria, produces the required acidity. Other pickles are made by placing vegetables in vinegar. Like the canning process, pickling (which includes fermentation) does not require that the food be completely sterile before it is sealed. The acidity or salinity of the solution, the temperature of fermentation, and the exclusion of oxygen determine which microorganisms dominate, and determine the flavor of the end product. When both salt concentration and temperature are low, Leuconostoc mesenteroides dominates, producing a mix of acids, alcohol, and aroma compounds. At higher temperatures Lactobacillus plantarum dominates, which produces primarily lactic acid. Many pickles start with Leuconostoc, and change to Lactobacillus with higher acidity.

1. Processing of Pickles with Vinegar

Vinegar is the one of the most important ingredients in pickles because it is the chief means of preserving them. Vinegar contains acetic acid which prevents the growth of bacteria, yeasts and moulds. Usually vinegar for pickles making is first seasoned with spices and can be stored for several months and can be used occasionally.

Preparation of spiced vinegar

Ingredients

1. Sugar-42 kg
2. Cloves-1 kg
3. Coriander seed-1 kg 170 g
4. Mustard seeds-1 kg 450 g
5. Cardamom seeds-137 g
6. Celery seeds-112 g
7. Ginger-340 g
8. Caraway seed-46 g
9. Vinegar-38 Lit

Method of preparation

1. Dissolve sugar in vinegar by heating
2. The spices are placed in linear bags are suspended in the vinegar and heated for 6 hrs under simmer heating and then cooled
3. The bags are removed and the vinegar is filled into casks for storage
4. Fruit pickles are usually preserved with spiced vinegar

A. Apple pickle with vinegar

1. Apples are peeled, cored and cut.
2. Now vinegar, sugar, spices are boiled for about 5 min and the apples are added and boiled.
3. Take out apples and keep in the jars.
4. Then vinegar and sugar are re-boiled to a syrupy consistency and pour in the jar which contains previously boiled apples.

B. Jack fruit pickle with vinegar

1. Tender and green jack fruits are selected and peeled and cut into ¾” thick slices.
2. Pack the slices in glazed Jar covered with 8% brine solution and gradually increase the strength of the brine by 2% daily till it reaches 15%.
3. Allow the slices in the jar for 8-10 days to soften them.
4. Drain off the brine water. Then add spiced vinegar and store it

2. Processing of Pickles with Salt

A. Mango avakaya with salt

1. Mango pieces-600 g
2. Salt-150-175 g
3. Chili powder-75-50 g
4. Mustard powder-175 g
5. Oil -150-200 g

Method of preparation

1. Mix mango pieces with salt and other masala powders thoroughly.
2. Cover the pickle with oil. Week after week mix the pickle and remix the pickle in 2 stages.
3. Then transfer the pickle contents in to clean and sterile wide mouthed Jars and close with the lid and preserve it.

B. Mango pickle with Salt

- Sliced raw mango-1 kg
- Common salt powder-300 g
- Red chilli powder-30 g
- Turmeric powder -30 g
- Fenugreek powder-30 g
- Pepper powder-30 g
- Sesame oil-100 g

Method of preparation

1. Wash, peel, and slice them into longitudinal slices
2. Immerse slices in 5% brine solution to prevent darkening due to exposure to air
3. Drain the brine, and add powdered salt, red chilli powder and other condiments powders and mix well
4. Transfer the contents to a wide mouth glazed jar and close with the lid
5. Add sesame oil and mix well and preserve it

3. Preservation of raw mango slices in brine/salt solution

More than any other fruit or vegetable, people preserve raw mango cut-pieces in brine/salt solution over a long period of time and utilize them for making pickles and chutneys as bulk. This is a large-scale demand and commercial practice among professional pickle makers, who by using such raw mango slices preserved in brine as and when needed, and by avoiding making pickles all at once in the season for year-long sales, save considerably on the inventory and investment costs of expensive condiments and spices and oil used in pickles.

Method of preparation

- Clean, wash and wipe the raw and firm mangoes
- Cut the mangoes into slices
- Fill the mango slices in a clean and sterile glass jar
- Prepare 10% conc. Salt/brine solution
- Pour the salt solution into glass jar filled with mango slices
- Keep a wooden disc over the jar so that the slices are fully immersed
- Since salt draws water from the mango slices salt should be added periodically to maintain the salt concentration at 10 per cent.
- Store the mango slices in brine solution for several weeks and can be utilize them for making pickles and chutneys as bulk when ever needed.

4. Processing of Oil-based pickles

Oil-based pickles containing one or more edible oils are highly popular in India. Mango, lime, cauliflower and turnip pickles are the most important oil pickles. Mango pickle is largely sold pickle in the country followed by cauliflower, onion, turnip and lime pickles. Cauliflower pickle is highly popular in Northern parts of India. For preparing the oil-based pickles, fruits or vegetables should be completely immersed in the edible oil. Oil pickles are generally highly spiced. The technology for the preparation of some of these pickles is given in this lesson.

Preparation of oil based mango pickle

The method of preparation of oil-based mango pickles varies in different parts of the country. The avakai pickle of Andhra Pradesh is a well known mango pickle in oil. It is very pungent and hot to taste. In north India, rapeseed /mustard oil is commonly used. But in south India, gingelly or sesame oil or groundnut oil are preferred. Hygienically prepared mango pickle should have a shelf life of about 1 to 2 years at room temperature. India, being the largest producer of mangoes, has great scope for export of mango pickles. Because of regional variations in taste, there are numerous recipe combinations of mango pickle.

Recipe for mango pickle

S.No.	Ingredients	Quantity (g)
1.	Mango slices	900
2.	Common salt, powdered	226
3.	Grounded Fenugreek (methi)	113
4.	Grounded Nigella (kalaunji)	28
5.	Turmeric powder	28
6.	Red chilli powder	28
7.	Black pepper	28
8.	Fennel or aniseed (saunf)	28

1. For making mango pickle, fully developed but under-ripe tart varieties of mango are selected.
2. They are washed with water; peeled or unpeeled and sliced longitudinally with a stainless steel knife.
3. The mango kernel stones obtained during slicing are discarded.
4. Sliced raw mangos are mixed with common salt to extract some of the moisture from the slices or immersed in brine solution of 2-3% strength to prevent blackening of the cut surface.
5. To the drained slices, a partially ground mixture of spices including onions and garlic are added.
6. Depending on the choice, availability and regional preferences the spices are selected and may include coriander, fenugreek seeds, nigella, fennel, cumin seeds, powdered turmeric and red chillies are added.
7. The whole admixture is filled into a clean glass or glazed jars and covered with the chosen edible oil i.e. mustard or gingelly or groundnut oil.
8. Table sugar is sometimes added in small amounts along with spices to produce a good taste blend.
9. Also, if the mango variety is not sour enough, a small amount of acetic acid is sometimes added.
10. Spices, oil and mango slices can be subjected to heating at low temperature for about 10-15 min. and add if texture of the slices are not softened. Such heat treatment reduces the initial bacterial load and would help in attaining a longer storage life of mango pickle.

11. Usually it takes 2-3 weeks for the completion of fermentation by natural micro biota.
12. Pickles usually have a shelf life of about slightly more than 1 year when prepared and packed in hygienic conditions.

Preparation of oil based Lime Pickle

Good quality lime are selected for making lime pickle and washed thoroughly before use.

1. The cleaned lime fruits are sliced and cut into four pieces.
2. About one-fourth of the cut lime fruit pieces are squeezed to extract juice.
3. Later, the remaining lime pieces are mixed with salt and extracted lime juice and spices such as turmeric, red chilli powder, cardamom, cumin, aniseed, black pepper, etc.
4. The mixture is filled into clean glass or glazed jars kept usually in sunlight for about a week. During this period the useful biochemical changes takes places.
5. At the end of one week, selected edible oil which is previously heated and cooled, is mixed with the pickle mixture and stored.

Preparation of oil based Aonla Pickle

Aonla, also called as amla or Indian Gooseberry, is a minor sub-tropical deciduous tree indigenous to Indian sub-continent. Aonla fruits are round, ribbed and pale green. It is divided into six segments through pale liner grooves. It is quite hard with a thin and translucent skin. The raw fruit, due to its high acidic nature and astringent taste, is unacceptable to consumers. Besides, aonla has been an important ingredient in chavanprash, an ayurvedic health tonic. Aonla fruits are rich in ascorbic acid and tannins. It contains 500-1500 mg of ascorbic acid per 100 g of pulp which is greater than that of guava, citrus and tomato fruits. Aonla has been used for pickle and preserve making since ages in India. Chakaiya is a popular cultivar or variety of aonla used for pickle preparation.

1. The principle of manufacture of aonla pickle is the same as that of mango pickle. For making pickle, aonla fruits of suitable size are selected and thoroughly washed with water.
2. The fruits are manually pricked and treated with salt and kept aside for few days.
3. Spices such as turmeric, nigella seeds, red chilli powder, fenugreek, headless cloves, are mixed

4. Oil as per recipe are mixed to the salt treated aonla fruits and filled in suitable glass jars.
5. The filled jars are kept for about a week in sunshine for the biochemical reactions to take place. Such pickle can be stored at room temperature and relished.

Table 1. Evaluation of pickles

Type of Pickle	Uniformity in mixing of ingredients	Sourness of the pickle	Salt Taste /color	Oiliness/ dryness	Spice mix flavor
1.					
2.					
3.					
4.					
5.					
6.					
7.					

Spoilage of Pickles

Different kinds of spoilages occur in pickles. Some of them irrespective of the type of the pickle are briefly described as follows:

Shrivelling

Shrivelling occurs when vegetables like cucumber are placed directly in a very strong solution of salt, sugar or vinegar. Use of weak solutions to start with and later gradually increasing their strength avoids such defect.

Bitter taste

Bitter taste in the pickle may result from use of strong vinegar or prolonged cooking of spices or addition of excess amounts of spices.

Blackening

Blackening defect in pickles may result from use of iron knives for cutting the fruits and vegetables and use of iron containers for processing. Further, it may also result from growth of specific molds that cause blackening.

Dull or faded products

Pickles become dull and faded either due to use of inferior raw material or due to insufficient curing.

Softness and slipperiness

This is a most common defect and results due to the action of bacteria. Insufficient covering with brine or use of weak brine or insufficient covering with oil are invariable causes for this defect. By using a brine of proper strength and by keeping the pickle well below the surface of the brine or oil, this kind of spoilage can be eliminated.

Scum formation

When vegetables are placed in the brine for curing, a white scum is invariably formed on the surface due to the growth of wild yeast. This scum retards the formation of lactic acid and helps the growth of putrefactive bacteria that causes the vegetables to become soft and slippery. Hence, it is essential to remove the scum as soon as it is formed. Addition of about 1% acetic acid helps to prevent the growth of wild yeast in the brine, without hindering the formation of lactic acid.

Cloudiness

This defect is usually found in case of onion and some other vegetables and occurs when vinegar become cloudy and turbid, thereby spoiling the appearance of the pack. This could be due to microbial activity or due to the use of inferior quality or possible chemical action between vinegar and impurities such as calcium, magnesium and iron compounds.

Table 2. Evaluation of stored pickles

Type of pickle	Acceptability of taste/flavor	Blackening	Scum formation	Softness and slipperiness	Cloudi-ness
1.					
2.					
3.					
4.					

18

Processing of Dairy Products

Aim: To learn the principles of processing and method of preparation of different types of dairy products and evaluate their characteristics

1. Processing and preparation of Yogurt

Yogurt is a fermented milk product that contains the characteristic bacterial cultures such as *Lactobacillus bulgaricus* and *Streptococcus thermophilus*. All yogurts must contain at least 8.25% solids not fat.

- Full fat yogurt must contain not less than 3.25% milk fat
- Low fat yogurt not more than 2% milk fat
- Non-fat yogurt contains less than 0.5% milk fat

Ingredients required

1. Milk **the type of milk used depends on the type of yogurt –**
   A. Whole milk for full fat yogurt
   B. Low fat milk for low fat yogurt
   C. Skim milk for non-fat yogurt
2. Cream **to adjust the fat content**
3. Non-fat dry milk **to adjust the solids content**
   - The solids content of yogurt is often adjusted above the 8.25% minimum to provide a better body and texture to the finished yogurt.
4. Stabilizers **used in yogurt are alginates (carageenan), gelatins, gums (locust bean, guar), pectins, and starch.**
   - Stabilizers may also be used in yogurt to improve the body and texture by increasing firmness, preventing separation of the whey (syneresis), and helping to keep the fruit uniformly mixed in the yogurt.
5. Sweeteners, flavors and fruit preparations **are used in yogurt to provide variety to the consumer.**

6. Bacterial Cultures

   - **The main (starter) cultures in yogurt are** *Lactobacillus bulgaricus* and *Streptococcus thermophilus*.
   - The function of the starter cultures is to ferment lactose (milk sugar) to produce lactic acid. The increase in lactic acid decreases pH and causes the milk to clot, or form the soft gel that is characteristic of yogurt. The fermentation of lactose also produces the flavor compounds that are characteristic of yogurt.
   - Other bacterial cultures known as probiotic cultures such as *Lactobacillus acidophilus*, *Lactobacillus subsp. casei*, and bifido-bacteria may be added to yogurt

Probiotic cultures benefit human health by improving lactose digestion, gastrointestinal function, and stimulating the immune system.

General yogurt processing steps

1. Adjust milk composition & blend ingredients

Milk composition may be adjusted to achieve the desired fat and solids content. Often dry milk is added to increase the amount of whey protein to provide a desirable texture. Ingredients such as stabilizers are added at this time.

2. Pasteurize milk

The milk mixture is pasteurized at 185°F (85°C) for 30 minutes or at 203°F (95°C) for 10 minutes. A high heat treatment is used to denature the whey (serum) proteins. This allows the proteins to form a more stable gel, which prevents separation of the water during storage. The high heat treatment also further reduces the number of spoilage organisms in the milk to provide a better environment for the starter cultures to grow. Yogurt is pasteurized before the starter cultures are added to ensure that the cultures remain active in the yogurt after fermentation to act as probiotics; if the yogurt is pasteurized after fermentation the cultures will be inactivated.

3. Homogenize

The blend is homogenized (2000 to 2500 psi) to mix all ingredients thoroughly and improve yogurt consistency.

4. Cool the milk

The milk is cooled to 108°F (42°C) to bring the yogurt to the ideal growth temperature for the starter culture.

5. Inoculate with starter cultures

The starter cultures are mixed into the cooled milk.

6. Hold

The milk is held at 108°F (42°C) until a pH 4.5 is reached. This allows the fermentation to progress to form a soft gel and the characteristic flavor of yogurt. This process can take several hours.

7. Cool

The yogurt is cooled to 7°C to stop the fermentation process.

8. Add fruit & flavors

Fruit and flavors are added at different steps depending on the type of yogurt. For set style yogurt the fruit is added in the bottom of the cup and then the inoculated yogurt is poured on top and the yogurt is fermented in the cup.

9. Package

The yogurt is pumped from the fermentation container/vat and packaged as desired.

Table 1. Evaluation of yoghurt

Type of yoghurt	Setting type	Blending of the ingredients	Stability property	Sourness	Sweetness
1.					
2.					
3.					
4.					
5.					
6.					

2. Processing and preparation of Butter milk

It is a popular refreshing drink prepared from the by-product produced during the preparation of butter/makkhan from dahi. Butter milk is also known as Mattha, Chhachh or Chhas in northern part of India. In south India it is called as Majjige or Majjika. The sweet variety of the product is relished in the northern part of the country, whereas the sour variety is preferred in the south.

Product description

Butter milk is similar to skim milk in composition except acidity of the product. It is generated during the production of butter from dahi by desi method. To this butter milk salt, coriander, ginger, onion are added to improve palatability. It is popular because of its aroma, mildly acidic taste developed during fermentation by mixed strains of lactic acid bacteria. In southern part of the country butter milk is prepared by churning high acid curd and addition of water to reduce acidity. Butter milk is consumed directly or along with rice and pickle.

Method of preparation

1. Milk is boiled and then cooled to 30° – 35°C.
2. It is added with dahi culture or previous day's dahi at the rate of 1.0 to 1.5%.
3. Milk is allowed to set overnight.
4. Set curd is stirred using a mathani (wooden/steel stirrer with impellers) driven by a small rope in to-and-fro circular motion. During this action, small grains of butter are formed and raised to top of the vessel and it is scooped out from time to time.
5. When all the butter is recovered, the residual watery fluid is referred as butter ilk/chhach/mattha/chhas.
6. This can be consumed directly or sweetened or salted and added with spices based on the preference.

3. Processing and preparation of Lassi

Lassi, similar to butter milk, is a refreshing summer beverage popular in north India. Lassi is a white to creamy white viscous liquid with a sweetish, rich aroma and mild to high acidic taste. It is flavored either with salt or sugar and other condiments, depending on regional preferences.

Product description

Lassi can be described as a fermented milk beverage obtained after the growth of selected lactic acid bacteria in heat treated milk followed by sweetening with sugar. It is consumed as a cold refreshing beverage in summer. It is prepared by breaking the curd in to fine particles by agitation, addition of sugar, water and optionally flavor.

Method of preparation

Lassi is becoming popular and attracting demand throughout the year. To meet the consumer demand many dairies have started producing lassi on commercial scale. Similar to dahi, fresh, good quality milk is essential for production of good quality lassi.

1. Raw milk is standardized for the fat content ranging between 1.5 – 3.8% and 9% SNF.
2. Standardized milk is heated to 90°C for 15min, cooled to 60°C and homogenized at 150kg.cm2 and 50kg/cm2 at 1st and 2nd stage respectively.
3. Milk is cooled to 30 -32°C , inoculated with lactic culture and incubated to attain the pH of 4.5.
4. The curd is broken with the help of a power driven agitator.
5. Sugar syrup (25% syrup) is added to the mix to give 12% sugar concentration in the blend.
6. Low methoxy pectin after making solution in water/syrup can also be added @ 0.5% at this stage as a stabilizer to improve the appearance and mouth feel.
7. The mixture can be flavored with rose water and homogenized to improve body and texture.
8. It is packed and stored at refrigeration temperature.

Packaging of butter milk and lassi

Butter milk and lassi are normally packed in polyethylene pouch. Form fill seal machines are widely used for high speed packing. Lassi and butter milk are filled in 60 – 80 micron LLDPE (linear low density polyethylene) pillow type pouches of 200 ml at the rate of 5000 pouches per hr. Lassi subjected to UHT processing is packed aseptically in rigid laminates.

Shelf life of butter milk and lassi

Butter milk and lassi packed in LLDPE pouch can be stored upto 7days at 5°C without any significant change in its sensory qualities. Lassi subjected to UHT treatment and packed aseptically has shelf life upto 120days.

The therapeutic value of fermented milk drinks is largely dependent upon the presence of live-active bacteria. UHT processing will destroy the lactic acid bacteria reducing the food value of the sterilized products.

Table 3. Evaluation of butter milk and Lassi

Name of product	Fermented/ non-fermented	Curdling particles/ or without curdling	Sourness	Sweetness	Consistency Thin/thick/ moderate	Off-flavors/or pleasant flavors
1.						
2.						
3.						
4.						

4. Preparation of Indigenous milk products

Cream

When milk is allowed to stand for some time, the fatty portion tends to collect on the top forming a layer called cream. Cream can be rapidly separated by applying centrifugal force. The equipment commonly used is called cream separator. The fat content of the cream may vary widely depending on the fat content of milk and method of separation.

Sour Cream

Sour cream is the one in which the acidity is more than 0.2 per cent expressed as lactic acid

Khoa: Khoa is prepared by evaporating whole milk in an open casr iron pan, with continuous stirring. khoa is used in the preparation of various types of milk sweets and is consumed as such with the addition of sugar.

Pannier: This is prepared by coagulation of milk by rennet. Milk is heated to about 70-80c for 10 minutes and cooled to about 37c. Lactic acid is then added at the rate of 20g per 100kg milk followed by rennet solution (20ml per 1000 kg of milk). The milk is well mixed and allowed to stand for 60 minutes. The resulting curd is transferred in small lots to clean bamboo baskets and the whey is collected. The collected whey is strained through muslin cloth. The cheese is preserved by allowing it to float in the whey. When allowed to ripen in the whey for 12-36 hours, the product has a firm body and smooth texture. Pannier is fried in fat and added to soups and vegetables

Channa (acid coagulated curd)

Milk is heated to boiling and cooled to above 50^{0}C. A dilute solution of citric acid is added with stirring till the milk coagulates. The coagulated milk is filtered through a cloth. The coagulated milk is called Channa. The Channa is used in the preparations of various milk sweets.

Casein manufacturing: The process of manufacture of protein from skim milk is as follows: Dilute hydrochloric acid (1 part of con HCI + 4parts of water) is added to skim milk with constant stirring till pH of the fluid reaches 4.6. After holding for about 30 minutes the whey is drained. The casein curd is washed twice by suspending in 6 times its volume of water allowing the curd to settle and draining off the supernatant fluid. The casein is taken in a cloth and the adhering water is pressed out in a press. The curd is transferred to trays in thin layers and dried in a tunnel drier at 125-130F (50-55C) in a current hot air. The dry casein is powdered in a hammer mill.

Evaporated milk

Evaporated milk (unsweetened condensed milk): The process of preparation of evaporated milk consists of the following steps: (1) Filtration and pasteurization, (2) Fat standardization, (3) Evaporation in vacuo, (4) Homogenization, (5) Filling in cans and sealing and (6) Sterilization at 240-245 degree F for 15 minutes. Evaporated milk is used for feeding infants. It contains about 2.25 times the solids of fresh whole milk and is generally fortified with vitamin D (800 I.U/kg of evaporated milk). The composition of evaporated milk in given in table

Condensed milk: Sweetened condensed milk is used mostly in the manufacture of cakes, ice cream and confectionary. The process of its preparation consists of the following steps: (1) Filtration and pasteurization of milk, (2) preheating and evaporation, (3) Addition of hot sterilized syrup of sucrose, (4) Homogenization and (5) Canning it.

Table 4. Evaluation of indigenous milk products

Type of indigenous milk product	Textural evaluation characteristics
Cream	
Sour Cream	
Khoa	
Pannier	
Channa	
Casein	

19

Preparation of Ready to Serve Beverages

Aim: To learn and prepare ready to serve beverages

1. Preparation of ready to serve beverage from Carrot with Sour-Orange Juices

Procedure: The RTS beverage was prepared by using freshly prepared carrot juice, sour orange juice, water and sugar as per FPO specification.

1. The juice of carrot and sour-orange were mixed in four different proportions i.e.100:0, 60:40, 50:50, and 40:60 respectively.

   T1 - RTS beverage with 100:0 carrot juice and sour orange juice

   T2 - RTS beverage with 60:40 carrot juice and sour orange juice

   T3 - RTS beverage with 50:50 carrot juice and sour orange juice

   T4 - RTS beverage with 40:60 carrot juice and sour orange juice

2. The requisite amount of sugar and citric acid were dissolved in water to prepare sugar syrup up to boiling stage, and strained through the muslin cloth and added after cooling, to get the desired TSS of 14° Brix -20° Brix.
3. The combined carrot and sour orange juices were then mixed with the sugar syrup in the correct quantity to get the RTS beverage.
4. All the ingredients were dissolved through homogenizing, and heated at 85°C for 20 minutes. It was removed from the fire and allowed to cool for 10 minutes. Subsequently 70 ppm Sodium metabisulphite (SMS) was added and mixed well with the solution.
5. Prepared RTS beverage was poured into pre-sterilized 200ml capacity glass bottles and capped with stopper immediately, leaving 1" head space on the top and closed with caps air tightly.
6. The sealed bottles were sterilized in hot water bath at 80°c for 30 minutes. The bottles were removed from the water bath and allowed to cool.

2. Preparation of ready to serve (RTS) Drink/Sweetened Juice

Aim: To prepare ready to serve drink from fruits.

Theory: Ready to serve (RTS) drink is a type of fruit beverage which contains at least 10% of fruit juice and 10 degree Brix total soluble solids (TSS) with 0.3% acidity. It is not diluted before serving, hence it is known as ready to serve beverage. Mango drink, guava drink, pineapple drink are the commercial products available in the market.

Sweetened fruit juice is categorized as ready to serve beverage in which the minimum juice content shall be 85% with a minimum TSS of 10%, while unsweetened juice is a natural juice having 100% fruit juice with natural total soluble solids. These are preserved by heat processing method. Sweetened apple, pear juices are commercially manufactured.

Raw material, ingredients, machinery required

1. Mango, orange, lemon, lime, litchi, pear, kiwi and apricot are used for making RTS.
2. Stainless steel knives, juicer/screw type juice extractor, utensils for cooking and mixing, glass bottles, sterilization tank, gas bhatti etc
3. Recipe for preparation of RTS from different fruits:

   Pulp/juice - 1.5 litre

   Sugar - 685 g

   Citric acid - 3.5g

   Water - 7.8 litre

   Potassium Meta-bisulphite (KMS) - 1.5g

   Note: RTS drink can also be preserved by adding not more than 70 ppm SO_2 or 120 ppm benzoic acid.

Procedure for preparation of ready to serve (RTS) drinks (Follow flowsheet Fig. 1 for preparation of RTS drink)

1. Take fruit pulp/juice and mix with syrup which is prepared by mixing sugar with water and citric acid.
2. Homogenize the mixture for proper mixing.
3. Heat the drink to boil for pasteurization.

4. Fill in the glass bottles (200ml capacity) while still hot.
5. Crown cork the bottles and process in boiling water for 20-25 minutes.
6. On cooling, label the bottles and store in cool and dry place.
7. Potassium meta-bisulphite (KMS) or benzoic acid can be added to the RTS drink as preservative.
8. Synthetic and approved artificial color and flavor can be added when declared on the label.

FPO specifications for Ready to serve drinks

	RTS drink
Juice/pulp content	Lime - Not less than 5% Other fruits - Not less than 10%
Total soluble solids (TSS)	Not less than 10%
Preservatives	Sulphur dioxide - Not more than 70 ppm Benzoic acid - Not more than 150 ppm
Synthetic sweetening agent	Not permitted
Color and flavor	Only approved artificial color and flavor should be used and mentioned on the label
Acidity	Not more than 3.5%
Incubation condition	No positive pressure/sign of bacterial growth when incubated at 37°C for one week

Evaluation of beverages

Beverage type	Amount of fresh fruits/ veg taken	Amount of pulp/ or juice taken	% of juice in the beverage	TSS%	°Brix	Yield of juice
1						
2						
3						
4						
5						

6

7

3. Preparation of Fruit Nectar

This is a type of fruit beverage containing at least 20% fruit juice/pulp and 15% total soluble solids besides about 0.3% acid. It is also not diluted before serving.

Method of preparation:

For the preparation of 10 lts. of fruit nectar, follow the procedure given below:

1. Calculate the amount of juice required as per commercial specification Required juice = (20/100) × 10 = 2.0 lts.
2. Measure the TSS using a refractometer (say the TSS is 30%) Calculate the total solids content of the juice i.e. 0.3 × 2.0 = 0.60 kg.

   The final required TSS content in the product is to be say 15%. The TSS required to be added to obtain the final product is (0.15 × 10 – 0.60) kg = 0.90 kg.
3. The amount of soluble solids in the form of citric acid and KMS is Citric acid @ 0.3%, in the final produce is 30 g i.e. 0.03 kg.
4. Amount of solids to be added in the form of sugar is 0.90–0.03=0.87kg.
5. Add calculated amount of sugar to about 2 lts of water and heat it till it dissolves completely. Add citric acid and juice to the sugar syrup and makeup the volume to 10 lts. Mix it well.
6. Heat the RTSB / Nectar up to 90°C, fill hot in clean pre-sterilized glass bottle up to brim, seal and cool in the air or Fill in bottle, seal and heat process (90°C for 25 min.).

4. Preparation of Cordial (Lime)

1. Extract the juice and strain through a fine muslin cloth to remove all pulp
2. Add preservative (KMS) 2 gms per litre of lime juice
3. Pour in bottles and keep for 1-2 months. All the sediment settle down and the juice becomes clear
4. Pour all clear juice without disturbing the sediment. Use the recipe and proceed as for RTS beverage.

5. Preparation of Sherbet

Sharbet should contain at least 65% total soluble solids, suitably acidified and may or may not contain fruit juice.

Method of preparation:

For the preparation of 10 lts. of sherbet follow the procedure given below:

1. Take about 6 lts. of water and add approx. 6.5 kg of sugar and dissolve it properly.
2. Add the required flavor.
3. Make up the volume to 10 lts. And adjust the TSS to 65%.

6. Preparation of Squash

Mango Squashes

To prepare following specifications 4 different Mango Squashes the amount of ingredients required are:

Ingredients Required	**25% fruit juice 45⁰ Brix 0.8% Acidity (Kgs)**	**33% fruit juice 45⁰ Brix 1.0% Acidity (Kgs)**	**25% fruit juice 65⁰ Brix 1.0% Acidity (Kgs)**	**33% fruit juice 65⁰ Brix 1.0% Acidity (Kgs)**
1. Mango pulp	50	50	50	50
2. Sugar	69.5	49.58	89.2	64.75
3. Citric acid	1.06	0.57	1.9	1.4
4. Water	79	49	59	34
5. Preservatives (potassium met-bisulphite)	0.2	0.15	0.2	0.15

Orange squashes

To prepare following specifications 4 different Orange Squashes the amount of ingredients required are

Ingredients Required	**25% fruit juice 45⁰ Brix 1.5% Acidity (Kgs)**	**33% fruit juice 45⁰ Brix 1.5% Acidity (Kgs)**	**25% fruit juice 65⁰ Brix 2.0% Acidity (Kgs)**	**33% fruit juice 65⁰ Brix 2.0% Acidity (Kgs)**
1. Orange juice	50	50	50	50
2. Sugar	82.5	60.6	121.45	89.55

3. Citric acid	2.65	1.55	3.65	2.65
4. Essence of orange	1.4	5.7	1.75	1.45
5. Water	63.9	36.6	23.55	6.35
6. Preservatives (potassium met-bisulphite)	0.2	0.15	0.2	0.15
7. Orange color	A pinch	A pinch	A pinch	A pinch

Table. evaluation other fruit products

Fruit product type	Juice % in final product	TSS % in final product	Setting of the products	Taste/flavor of the products
1				
2				
3				
4				
5				

FPO specifications for Fruit beverages

Type of beverage	Minimum % of TSS (total soluble solids) in final product	Minimum % of fruit juice in final product
1. Fruit squash	40	25
2. Fruit syrup	65	25
3. Cordial	30	25
4. Sweetened juice	10	85
5. Fruit juice concentrates	32	100

Permissible limits of preservatives in fruit beverages

Fruit beverage	Preservative	Max. level permitted (ppm)
1. Fruit juice concentrates	Sulfur dioxide	1500
2. Squashes fruit syrups, cordials, fruit juices	Sulfur dioxide Or	350
	Benzoic acid	600

References

Afreen S.M.M.S. Premakumar K., Inthujaa Y. (2016). "Preparation of Ready-to-Serve (RTS) Beverage from Carrot with Sour-Orange Juices. International Journal of Innovative Research in Science, Engineering and Technology; Vol. 5, Issue 2.

American Meat Science Association 1997-2016

Avantina Sharma (2019). Text book of food science and technology, 3rd edition; CBS Publishers and Distributors Pvt. Ltd.

Encyclopedia of Dairy Sciences (Second Edition), 2011.

Fellows P.J. (2000). Food processing technology-principles & practice 2nd edition; Wood Head Publishing,CRC Press LLC.

Hui Y.H. (2014). Dairy science and technology hand book-principles and properties, Volume-1; John Wiley (WSE).

Jhari Sahoo and Manish Kumar Chatli (2015). Text book on Meat Poultry and Fish Technology Astral Publishers.

Kendall P., DiPersio P. and Sofos J. Drying Vegetables, Food and Nutrition Series No. 9.308.

Margaret McWilliams (2012). Foods-experimental perspectives, 7th edition; Prentice Hall, Inc.

Marjorie P. Penfield, Ada Marie Campbell (1990). In Experimental Food Science (Third Edition).

McGee, Harold (2004). On Food and Cooking: The Science and Lore of the Kitchen. New York: Scribner, pp. 291–296. ISBN 0-684-80001-2.

Melese Temesgen and Senayit Yetneberk (2015). "Effect of Application of Stabilizers on Gelation and Synersis in Yoghurt". Food Science and Quality Management Vol. 37: 90.

Pickle Bill Fact Sheet". 13 March 2008. Archived from the original on 13 March 2008. Retrieved 15 February 2018.

Potter (2007). Food Science 5ed by CBS Publishers.

Preeti Birwal, Megh R. Goyal and Monika Sharma (2021). Handbook of research on food processing and preservation technologies Volume 2: Nonthermal Food Preservation and Novel Processing Strategies; Apple Academic Press.

Processing of Horticultural Crops March 2012 "preparation of ready to serve beverages TNAU.

Ramandeep Kaur1, Gurpreet Kaur1, Rima1, Santosh Kumar Mishra-1, Harsh Panwar1, K.K. Mishra2 and Gurvir Singh Brar (2017). Yogurt: A Nature's Wonder for Mankind Intl. J. Food. Ferment. 6(1): 57-69, June 2017.

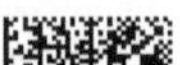

Rhee M.S., Lee S.Y., Dougherty R.H., Kang D.H. (2003). "Antimicrobial effects of mustard flour and acetic acid against Escherichia coli O157:H7, Listeria monocytogenes, and Salmonella enterica serovar Typhimurium". Appl Environ Microbiol. 69 (5): 2959-63.

Shafiur Rahman M. (2007). Hand book of food preservation 2nd edition; CRC Press.

Staff M.C. (1998). Cultured milk and fresh cheeses. pp. 124-125. In: R. Early (ed.). The Technology of Dairy Products. Second Edition, Blackie Academic and Professional, London, UK.

Swaminathan M. (1987). Food science chemistry and experimental foods. The Bangalore printing and publishing co.ltd, Bangalore.

Tamime A.Y. and R.K. Robinson (1999). Yoghurt: Science and Technology, CRC Press, Boca Raton, USA.

Vickie A. Vaclavik, Elizabeth W. Christian, Tad Campbell (2020). Essentials of food science food science text series): Springer.

Walstra, P. Wouters, J.T.M. and Geurts, T.J. (2006c). Chapter 22 Fermented Milks. In Dairy Science and Technology; Taylor & Francis Group, LLC: Boca Raton, FL, USA, pp. 551–573.

Walstra P., Wouters J.T.M. and Geurts T.J. (2006b). Chapter 11 Cooling and freezing. In Dairy Science and Technology; Taylor & Francis Group, LLC: Boca Raton, FL, USA, p. 297-307.

Yoghurt Production Staff (1998), Tamime and Robinson (1999), Walstra *et al.* (1999) and the website by Goff, www.foodsci.uoguelph.ca/dairyedu/yogurt.html.